TOP S
совершенн

KGB TRAIN

WORKING WITH AGENTS

РАБОТА С АГЕНТУРОЙ

1970

Translated from the original Russian
by Major Christoph P. Schwanitz (Ret.)

Conflict Research Group
London, 2025

Original edition published for internal use by the Ministry for State Security (KGB) of the Union of Soviet Socialist Republics.

This English language translation published by Conflict Research Group, London, United Kingdom, 2025

Copyright Notice

About the Conflict Research Group

Conflict Research Group (CRG) is a non-profit think-tank based in the United Kingdom, dedicated to advancing understanding of the art and science of Unconventional Warfare. With a focus on the academic study of guerrilla warfare, revolutionary warfare, asymmetric warfare, Fourth Generation Warfare, Fifth Generation Warfare and political unrest, CRG's work sheds light on the complexities and nuances of modern conflicts. By bringing critical and key works back into print, the organization serves as a vital resource for academics, policymakers, and military professionals seeking in-depth knowledge in these specialized fields.

At the heart of CRG's mission is the belief that a comprehensive understanding of Unconventional Warfare is essential for addressing contemporary security challenges. The group's research and publications delve into historical and contemporary case studies, exploring the strategies, tactics, and implications of irregular warfare. Through this rigorous analysis, CRG contributes to the development of more effective and adaptable strategies for dealing with non-traditional threats.

One of the key aspects of CRG's work is its publishing arm, which is dedicated to bringing into print seminal works on Unconventional Warfare. The group's publications cover a wide range of topics, from historical accounts of guerrilla movements to theoretical analyses of contemporary conflict dynamics and of course reprints of historical official publications. By making these works accessible to a broader audience, CRG aims to enrich the discourse on Unconventional Warfare and contribute to the development of more nuanced and effective approaches to resolving conflicts and disrupting, degrading and defeating unconventional threats.

CRG's research is categorised by its interdisciplinary approach, drawing on insights from military history, political science, sociology, and international relations. This holistic perspective allows the organization to address the multifaceted nature of unconventional warfare, considering not only military tactics, but also the granularity of the political, social, and economic dimensions of conflicts. Through this comprehensive approach, CRG provides a deeper understanding of the root causes and long-term implications of irregular warfare.

Publisher's Note

This English translation of *Working With Agents* (1970) is based on a Soviet-era KGB manual that was never declassified. The original document, produced by the KGB's First Chief Directorate, was intended for internal use in intelligence training and operational instruction. While the Soviet Union no longer exists, this text serves as an important historical resource on Cold War espionage tradecraft.

The translation presented here is the result of meticulous effort to ensure accuracy, clarity, and readability for an English-speaking audience. It includes refinements, contextual clarity, and structural modifications to make the material accessible without altering the meaning of the original text. Additionally, this edition features a *Translator's Note*, *Editor's Introduction* and an explanatory chapter about the KGB to provide historical and operational context.

Neither the translator or the publisher endorse or promote espionage activities but present this work purely for educational, historical, and academic research purposes.

Translator's Note

I grew up in East Germany during the 1970s and 80s at the height of the Cold War, with an uncle who was a high-ranking official in the notorious and now thankfully defunct East German Ministerium für Staatssicherheit or Ministry for State Security, more commonly known today as the Stasi.

I did not know it at the time, but before moving into his highly prestigious and important role as Stasi chief for East Berlin, my uncle had spent a large part of his career in the Stasi's Hauptverwaltung Aufklärung section (HVA) which dealt with foreign intelligence operations. This was the GDR's equivalent of the Soviet KGB or the American CIA or the British MI6.

As a young man, particularly during that chaotic period immediately following German Reunification, this all conspired to give me a deep interest in the intricacies of intelligence operations and ensured that I would later seek to forge a career in the field of intelligence. Of course, I would go on to do just that, serving with several different operational units within the German Federal Army until my retirement from the Bundeswehr in 2017.

As someone who had spent decades studying Russian language, culture, military and intelligence structures in the course of my work as an officer in the Army, I knew that translating these manuals would require more than just basic linguistic proficiency. It would demand an intimate knowledge of the nuances of each language, as well as a deep understanding of the cultural, historical, technical and operational contexts in which they were written.

One of the more significant challenges I faced, and the one which is least likely to be of any great interest to a reader of this book, was navigating the complexities of Russian grammar and syntax. Unlike German, which is known for its strict rules and conventions, Russian has a more relaxed system that can make it difficult at times to accurately convey meaning. For example, Russian word order often prioritizes grammatical function over semantic content, making it essential to carefully consider the context in which each sentence appears.

Furthermore, Russian relies heavily on prepositions and case endings to convey subtle shades of meaning, whereas German tends to rely more on verb conjugation and adverbial phrases. This meant that I had to be particularly mindful when translating individual words or phrases, as their meanings could shift significantly depending on the surrounding context.

Another significant challenge was capturing some of the more obscure technical jargon used in these manuals. With many of the KGB manuals in the cache dating to the 1960s and 70s, some of the old Soviet terminology has become obsolete and has been replaced by other terms within the Russian Federation intelligence services. The manuals display a wide array of obsolete and specialized terms for various aspects of intelligence operations, from types of agents to counter surveillance techniques to clandestine communications methods.

As someone who is very familiar with German and other NATO partners' intelligence operations and with their jargon and acronyms, I found myself constantly referencing my own knowledge base to ensure that I accurately conveyed the intended meaning of any given passage that included technical operational language. In those few cases where I could not be 100% sure of a technical term's meaning, I simply extrapolated to the best of my ability.

The fact that many of these manuals remain classified in Russia even today speaks volumes about their significance and relevance. It's likely that some are still being used by Russian Federation intelligence services to train new personnel, while others would have no doubt been declared obsolete but remain sensitive due to the nature of their contents.

I have a responsibility to ensure that these manuals are translated accurately and with appropriate sensitivity. It's not just about conveying technical information; it's also about respecting the cultural and the operational context in which they were written. In many ways, translating these KGB tradecraft training manuals was akin to conducting an archaeological excavation into the past. Each sentence or phrase revealed a piece of history that had been hidden away for decades, waiting to be uncovered and shared with the world.

As someone who has spent years studying Russian language and culture as well as evaluating the potential threats which an adversarial Russian Federation might in the future pose to my homeland and our NATO partners, I'm proud to have played a role in making this significant historical material available for public consumption.

I would like to thank "DC" and "CB" from Conflict Research Group for assigning me the delicate but critical task of translating this important material. Having become well informed of the vital work being undertaken by Conflict Research Group, I am honoured to be of service even in this small way.

I would like to thank my beloved wife, Birgitt, for dealing with my many absences and long days spent locked away in my study working on this material and accepting it all with grace and good humour.

I would like to also thank Birgitt for her assistance in helping me translate certain more complex passages from German to English and for proof-reading the final manuscript to correct my abysmal English language grammar. As always, without her, I would be diminished.

Please note that any errors or omissions in these translated pages which may serve to detract from the original Russian language documents are mine and mine alone.

Christoph P. Schwanitz,
Major, KSA (ret.)
Görlitz, 2025

About the KGB

The KGB was the foreign intelligence and domestic security agency of the Soviet Union. It was established on the 13th of March, 1954, soon after the death of Soviet dictator Josef Stalin and it was dissolved with the fall of the Soviet Union on the 3rd of December 1991. The KGB's First Main Directorate was split off and became the Russian Federation's current foreign intelligence service, the SVR.

In addition to its primary responsibilities for foreign intelligence and domestic counterintelligence, during the Soviet era the KGB also had duties such as safeguarding the country's political leadership, overseeing border troops, and carrying out surveillance of the population.

In this book, we are dealing solely with the foreign intelligence aspects of KGB operations, so we shall look at the KGB's foreign intelligence apparatus.

The KGB's First Main Directorate, also known as the First Chief Directorate, was responsible for intelligence operations outside of the Soviet Union.

The directorate was organised into various directorates, including:

Directorate "R" -   Planning and Analysis,
Directorate "S" -   Illegals,
Directorate"T" -   Scientific and Technical Intelligence,
Directorate "K" -   Counter-Intelligence,

Directorate "OT" -   Operational and Technical Services,
Directorate "I" -   Computers Service (would
                      be known as "IT" today),
Directorate "A" -   Active Measures,
Directorate "RT" -   Operations within the USSR

In addition to the administrative directorates listed above, the First Main Directorate had various "Desks" or "Departments" dedicated to operations in various parts of the world or other specialised functions. These were:

1st Department -   North America
2nd Department -   Latin America
3rd Department -   UK, Australia, NZ, Scandinavia,
                      Malta
4th Department -   East Germany, Austria, West
                      Germany
5th Department -   France, Spain, Portugal,
                      Luxembourg, Switzerland, Greece,
                      Italy, Yugoslavia, Albania, Romania
6th Department -   China, Laos, Viet Nam, Cambodia,
                      North Korea, South Korea
7th Department -   Thailand, Indonesia, Malaysia,
                      Singapore, Japan, Philippines
8th Department -   Afghanistan, Turkey, Iran, Israel
9th Department -   English-speaking countries in
                      Africa (South Africa, Rhodesia/
                      Zimbabwe, Tanzania, Nigeria, etc.)
10th Department -   French-Speaking Countries in Africa
11th Department -   Liaison with other communist
                      countries' intelligence services
                      particularly Cuban and Warsaw Pact
                      nations (was previously known as
                      the "Advisor's Department")
12th Department -   Covers
13th Department -   Covert Communications
14th Department -   Forgeries
15th Department -   Operational files and archives

16th Department -	Signals intelligence
17th Department -	India, Pakistan, Bangladesh, Sri Lanka, Burma, Nepal
18th Department -	Egypt, Syria, Libya, Iraq, Oman, Saudi Arabia, Kuwait, Sudan, Jordan, Morocco, United Arab Emirates/Trucial States
19th Department -	Soviet Expatriates and Emigres
20th Department -	Liaison with 3rd World / newly independent states

It is the KGB's First Main Directorate which was the publisher of these manuals and is most likely that they were produced by staff of the First Main Directorate's *Directorate OT*, which was responsible for Operational and Technical Support functions.

KGB foreign intelligence networks were operated by a KGB Residence or *Rezidentura* as it is known in phonetic Russian. Please note that in these translations, we sometimes refer to the Residencies using the equivalent CIA term "Station". This is simply to reduce the possibility of confusion and to differentiate between a Residence and a private residence such as those used as safe houses or clandestine postal addresses. Similarly, within the translation in these situations, we will refer to the KGB Residence's "Resident" using the CIA term "Station Chief".

The KGB Resident or Station Chief was a legal intelligence officer usually operating under diplomatic cover as a "cultural attache" or similar. Diplomatic credentials gave the Resident diplomatic immunity meaning the security forces of the country in which he was operating could never arrest a KGB Resident. At best they could have him expelled from the country like any other diplomat, but this usually had serious diplomatic consequences. Instead, most countries usually worked out fairly quickly who was KGB within their local Soviet embassy and they usually allowed the KGB Resident to operate, but placed him and other Soviet embassy staff under heavy counterintelligence surveillance.

Typically, a KGB Residency was organised into different sections or "lines". Each section had a separate function which supported operations conducted out of any given Residency. These sections could be further categorised into separate functions - Operational and Support.

Operational sections of a KGB Residency were as follows:

Section "EM" -	Intelligence and surveillance of the activities of Soviet Emigres in the host country
Section "KR" -	Counterintelligence and protective security of the Residency
Section "N" -	Support to "illegal" Intelligence Officers in the host country
Section "PR" -	Economic, military, political intelligence on the host country or region as well as active measures such as black propaganda
Section "SK" -	Surveillance and reporting on Soviet diplomatic staff in the host nation.
Section "X" -	Technical intelligence and advanced technology acquisition and transfer.

Support sections of a KGB Residency were as follows:

Section "OT" -	Technical support
Section "RP" -	Signals intelligence
Section "I" -	Information technology

Support staff not assigned to their own specific section included drivers, signals operators, cipher clerks, administrative staff, finance personnel.

Table of Contents

PART III:
TRAINING AND GUIDANCE OF AGENT
OPERATIONS

PART IV:
VETTING AND ONGOING ASSESSMENT
OF AGENTS

PART V:
INTERACTION BETWEEN INTELLIGENCE
OFFICERS AND AGENTS

PART VI:
METHODS OF TERMINATING AGENTS

Editor's Introduction

The original Russian language manual this English translation is based on was found on the deep web in a cache of scanned older Soviet KGB training materials in a folder on a Russian language .onion site. It is believed that these materials were posted by a dissident many years ago, perhaps even as long ago as 2010 or 2012 based on file metadata. The cache was later posted on the surface web, where to this day scans of the original Russian language documents can still be found through a simple search on any search engine.

Various think-tanks from English-speaking countries had made promises to translate and publish these materials, but despite waiting over five years for them to do so, no apparent progress has been made. With the Russian invasion of Ukraine in February 2022, it appears that translation and publication of the KGB training manuals is no longer a priority for these organizations. As a consequence, and with no clear end to the Ukraine War in sight at the time of writing, we have gone ahead and translated and published the KGB manuals from the cache ourselves.

Please note that we are not the first to publish English language translations of some of these materials. Circa 2020, enterprising persons unknown, in a blatant cash-grab, ran a couple of these documents through some translation software, probably Google, before dumping the resulting unedited text into a book format for publishing on Amazon. We purchased a copy of each of these translations to see whether there would still be a requirement for our professionally translated editions. Sadly, all were largely unreadable, therefore, we pushed ahead with our project.

As Conflict Research Group deals mostly with unconventional warfare, resistance, and inform/influence operations from the perspective of non-state actors, it would seem to the casual reader that espionage training materials from a former nation-state intelligence agency such as the Soviet KGB would fall well outside our remit.

This is simply not the case. During the Cold War, the Soviet Union, its Warsaw Pact satellites and other communist states such as the People's Republic of China and the Democratic People's Republic of Korea invested many billions of dollars in to supporting subversive and revolutionary groups fighting against western interests from Southeast Asia to the Middle east, to Latin America, to Southern Africa. Soviet support for such groups was not limited to weaponry and war materiel, but also included training in communist political theory, revolutionary and guerrilla warfare and of course, in clandestine tradecraft to allow members of a revolutionary or terrorist group to organize, plan and conduct their activities in secret.

Western-trained security forces typically used extremely effective British, French or American counter-intelligence and counter-insurgency methods to detect and destroy insurgent undergrounds or espionage rings at or before their nascent stage, so there was a requirement for guerrilla or terrorist groups sponsored by the Soviet Union to be given the most effective tradecraft training available in the communist world, and that came from the KGB's First Main Directorate.

Two English language resources which closely follow KGB procedures and concepts can be found in the 1980s-era South African Communist Party pamphlet *How to Master Secret Work* and in the 1970s-era document *Security and the Cadre* produced by a Puerto Rican separatist group operating in the US, the Fuerzas Armadas de Liberacion Nacional (FALN). Anyone reading through those two sources and then reading one of these KGB manuals will soon find examples which appear in all three, sometimes almost word-for-word.

Unlike some Western intelligence services such as the CIA, which train personnel in very specific, complex tradecraft techniques and methodology (some involving literal magician's sleight of hand), the KGB instead concentrated on teaching its personnel general concepts. This forced the KGB operative to become highly adaptable and imaginative in putting those concepts into action in the field. This lack of a "toolkit" of relied-upon tactics, techniques and procedures meant no clear patterns were set, making it just that much harder for western counterintelligence services to anticipate the specificities of a KGB intelligence officer or agent's tradecraft in the field.

In closing, I would like to thank Major Chris Schwanitz for his accurate translations of these materials, as well as for standing by for almost 18 months while we decided whether or not to go ahead with this project. I would also like to thank "DC" and her OSINT team for backtracking the circumstances of how the original scanned documents came to be posted online. Finally, I would like to thank you, the reader, for your interest in this project and for your support of CRG by purchasing this book.

CB
London, 2025

PART I

PRINCIPLES OF WORKING WITH AGENTS

Introduction

Agents constitute the principal instrument of intelligence operations. The success of all clandestine activity is directly contingent upon the quality and rigor of agent handling. The primary objectives in managing the agent network are: the cultivation of ideological loyalty to the socialist state; the development of political consciousness; the formation of disciplined and capable executors of operational assignments; and the full exploitation of each agent's intelligence potential.

The process of ideological and political education, as well as the systematic instruction of agents in intelligence tradecraft, constitutes a continuous obligation of the intelligence services of the socialist state. This process encompasses all members of the agent network, including those considered ideologically steadfast and operationally proficient.

The necessity of this ongoing ideological engagement stems from the operational reality that agents often reside within or operate across capitalist societies. Even those individuals aligned with the socialist cause are subject to the influence of bourgeois propaganda. Such exposure may distort their understanding of international developments or the policies of fraternal socialist states.

Moreover, due to the nature of their assignments, agents frequently operate within hostile environments. Persistent exposure to adversarial influence carries the risk of psychological degradation, political disorientation, or operational compromise.

Newly recruited agents, in particular, typically lack experience in clandestine activity. They may not fully comprehend the critical importance of strict adherence to security protocols, especially concerning the maintenance of operational secrecy.

Accordingly, the intelligence service must provide systematic instruction in clandestine technique to ensure the agent fully grasps the necessity of tradecraft. Intelligence aims to exploit each agent's capabilities in the most operationally effective and expedient manner. Agents are to be thoroughly trained in methods of intelligence acquisition and in the execution of specific assignments. Concurrently, the service must endeavor to strengthen and expand the operational capacity of the agent network.

Work with agents includes active operational management, continuous observation, assessment of development, and regular verification of loyalty and effectiveness. These elements: direction, study, and vetting form an indivisible cycle of continuous operational control.

Operational experience within the foreign intelligence services of socialist states affirms the following foundational principles as the bedrock of agent network management:

- Ideological Commitment
- Purposefulness and Concreteness
- Individualized Approach
- Ongoing Vetting and Assessment
- Operational Security and Tradecraft

1. Ideological Commitment of Agents

Socialist intelligence services place paramount importance on the sustained ideological cultivation and political education of agents. Agents must be raised in a spirit of progressive consciousness, instilled with loyalty to the socialist homeland, and firmly convinced of the historical inevitability and correctness of Marxist-Leninist doctrine.

This process of ideological formation is complex and labor-intensive, requiring deliberate and persistent effort by intelligence officers. It takes place within the context of continuous ideological confrontation, where agents are exposed to the corrosive influence of bourgeois ideology seeking to erode political conviction and psychological resilience.

Agents who are not subjected to consistent ideological engagement are demonstrably more vulnerable to hostile influence. During periods of acute political crisis, such agents may become disoriented, leading to political confusion, operational hesitancy, disengagement, or outright defection.

Operational experience has repeatedly shown that a lack of ideological attention results in negligent fulfillment of assignments, compromise of the agent, and betrayal of the service.

Therefore, it is essential to assess the agent's political disposition continually, to identify signs of doubt, and to apply targeted ideological reinforcement. The ideological education of agents is bolstered by the entire trajectory of historical development, which affirms the correctness of Marxist-Leninist doctrine and the tangible achievements of the socialist camp, particularly in its struggle against imperialism and for global peace and security.

By participating in this global struggle, agents internalize their political role, reinforcing their ideological formation. The intelligence service emphasizes the true nature of the bourgeois state, as an instrument of the monopolist bourgeoisie for repressing the working class and subjugating entire nations, and instills in its agents the understanding that by fulfilling their operational assignments, they act not against their own people but against their exploiters.

Unlike the intelligence services of capitalist states, which serve as instruments of domestic repression and imperialist conquest, employing ideological subversion, deception, blackmail, and coercion, socialist intelligence services engage with agents on the basis of ideological affinity and political consciousness.

The psychological dimension must also be considered. Cooperation with the intelligence service may involve hesitation or inner conflict, even for ideologically sympathetic individuals. Doubts may arise during recruitment or operational activity due to distorted portrayals of intelligence work in capitalist propaganda, which seeks to discredit intelligence officers as immoral or dishonorable. Overcoming these misconceptions is a core objective of ideological-political education within the agent network.

2. Purposefulness and Concreteness

All operational activity of the socialist intelligence service is governed by strict purposefulness. Each agent is assigned a specific operational direction corresponding to their capabilities. The agent must clearly understand the nature of their assignment, what is to be done, and how best to achieve it.

Concrete guidance implies more than assigning objectives and demanding results. The service must study the conditions under which the task is to be executed, collaborate with the agent in planning methods of execution, provide detailed instructions, and supervise implementation. This approach enhances the agent's mastery of tradecraft, protects them from exposure, improves situational awareness, and ensures successful task fulfillment. It also facilitates assessment and vetting.

3. Individualized Approach

The principle of individualized handling requires that the following factors be considered in all agent relationships: basis of recruitment, political beliefs, reliability, tradecraft experience, social and professional standing, education, and personal character.

Intelligence must maintain comprehensive knowledge of the agent's personal circumstances. Case officers should possess detailed biographical information, including family, associates, workplace, access to classified materials, document storage habits, work schedule, travel patterns, and frequented locations.

A differentiated, individualized approach must be combined with strict professional objectivity, attentiveness, and empathy. This enables the development of a stable professional rapport, which is essential for sustained operational success.

4. Ongoing Vetting and Assessment of Agents

Regardless of recruitment basis, trust level, or duration of cooperation, the intelligence service must maintain uninterrupted observation and vetting of all agents.

This requirement arises from the reality that an agent's operational viability is subject to change due to developments in professional, social, or familial circumstances. Political views and attitudes toward cooperation may also evolve.

These changes can directly impact operational performance. Without timely recognition, effective agent management becomes impossible. Early identification allows for appropriate corrective action, whether releasing compromised agents, neutralizing double agents, or decommissioning those whose utility has expired.

5. Operational Security and Tradecraft in Agent Handling

Operational security is vital in all agent operations. Adherence to strict tradecraft ensures protection against infiltration, enhances agent safety, and secures successful mission execution.

Core tradecraft principles include:

- Knowledge of agent affiliation is limited strictly to officers with operational need-to-know
- No classified information is disclosed to agents unless operationally required
- All communication with agents is conducted via secure, need-to-know channels
- The identity of sources must not be revealed in reports or during agent contact
- Case officers must prevent careless actions that could expose agents, and must train agents in appropriate operational conduct

These principles, derived from the practical experience of

the KGB and its fraternal services, represent the codified doctrine for handling the agent network of a socialist intelligence service. Their consistent application ensures the long-term effectiveness, security, and ideological integrity of clandestine operations.

PART II

IDEOLOGICAL AND POLITICAL EDUCATION OF AGENTS

Introduction

The ideological and political education of agents constitutes one of the most critical tasks in the management of the agent network. To educate an agent ideologically means to persuade him, through concrete examples drawn from the current international situation, that the operations of the intelligence services of the socialist state are directed toward the struggle against imperialism, that they contribute to the strengthening of the global socialist system, and that they serve the cause of peace.

The better an agent understands this, the more clearly he will perceive the purpose of his collaboration with the intelligence service, and the more firmly he will link that collaboration to the broader objectives of the struggle against imperialism. In turn, this will make him more conscientious and reliable in the execution of his assigned tasks.

Such an agent views his cooperation with the intelligence service of the socialist state as a meaningful contribution to the active struggle against imperialism. He will more effectively expose the plans of imperialist states, influence public opinion, and penetrate government circles in support of the interests of the socialist cause.

Agents who are ideologically aligned with us, who understand the political nature of their mission, are more dependable and productive. The higher the agent's level of ideological-political development, the more reliable he becomes, the more fruitful his work will be, and the more precisely he will carry out the assigned intelligence tasks.

Accordingly, the intelligence services of socialist states must, as a rule, strive to cultivate ideologically steadfast and politically committed agents, individuals capable of conducting intelligence work under the most demanding conditions.

The history of socialist intelligence services provides multiple examples in which, through the persistent ideological-political work of intelligence officers, even individuals originating from hostile class backgrounds were won over to communist ideals and began to conscientiously carry out intelligence assignments.

Naturally, the degree of ideological commitment and political conviction among agents varies considerably. Some become firm adherents of the communist worldview. Others, while not fully embracing communist ideals, nonetheless support the foreign policy objectives of the socialist states, especially on questions of anti-imperialism and national liberation, and cooperate with intelligence in furtherance of those aims.

The foundation of ideological-political education within the agent network of socialist intelligence is the doctrine of Marxism-Leninism.

Party resolutions, state directives, and the speeches of socialist leaders equip intelligence services with necessary arguments and persuasive frameworks for addressing contemporary political questions. Of particular significance are the strategic conclusions regarding anti-imperialist struggle in the modern era. These are well summarized in the materials of the International Conference of Communist and Workers' Parties held in 1969, which intelligence officers are expected to study closely.

The documents of that Conference provide a class-based analysis of global developments, highlight emerging trends in the politics and economics of imperialist states, describe the growing aggressiveness of imperialism, outline the principal fronts of anti-imperialist resistance, and propose a comprehensive program for combating imperialist forces, along with the strategic goals of the international communist movement.

This program defines the main directions of the anti-imperialist struggle while allowing tactical flexibility and a broad range of methods suited to specific operational conditions. As a unifying document for communist and workers' parties, it provides a vital foundation for the ideological orientation of our intelligence work and of those agents engaged in the broader class struggle.

The principles set forth by the 1969 International Conference of Communist and Workers' Parties are not limited to a narrow partisan perspective. Rather, they are addressed to all those who stand in opposition to imperialism, to all who are prepared to fight for peace, freedom, and progress.

The Conference's foundational document, "The Tasks of the Struggle Against Imperialism at the Present Stage and the United Action of Communist and Workers' Parties," underscores that the principal trajectory of global development continues to be defined by revolution and socialism, by national liberation movements. The Conference reaffirmed that the decisive force in the anti-imperialist struggle is the global socialist system. Further successes of this system, and of its cohesion, are critical for defeating the old world and constructing the new.

Concern for strengthening the global socialist system is simultaneously concern for the development of the world revolutionary process and for waging an effective struggle against imperialism. These strategic principles, formulated by the 1969 Conference, provide a powerful weapon in the hands of the intelligence services of socialist states.

A key condition for successful ideological-political education of agents is the high level of ideological-theoretical preparation, party-mindedness, and principled character of the intelligence officer. This imposes a duty on every officer to continuously enhance their theoretical grounding and general educational level.

Political education of agents may take various forms, but the most effective and widely applicable is the personal ideological conversation, a direct exchange between officer and agent on political topics. These discussions may be occasioned by current events, or framed around literature, art, science, technology, or contemporary global affairs.

Such conversations, even when informal, must always maintain an educational tone and be conducted in a manner that does not appear forced or artificial. They must not give the impression that the agent is being "processed" or indoctrinated. The most effective approach is a sincere, dynamic, and naturally flowing exchange.

A political conversation between the case officer and the agent must be conducted in such a way as to awaken the agent's interest in the topic and guide him logically to correct conclusions. Whenever possible, it is advisable to link the subject of the conversation to the agent's operational duties, using specific, real-life examples drawn from his work or surrounding conditions. Intelligence information acquired by the agent can serve as persuasive material in ideological engagement: to dismantle enemy propaganda, unmask the lies of the bourgeois press, reveal the hidden machinations of imperialist actors, and expose the subversive activities of hostile political groups in the target country.

Each political conversation aims at a single overarching goal: to consolidate the agent's progressive worldview, broaden his understanding of political, economic, and cultural issues, and enhance his awareness of international developments. In such exchanges, the case officer must unobtrusively introduce Marxist-Leninist analysis of political events, societal dynamics, and international trends.

Personal discussions require advance preparation: defining the topic, planning the conversation, and tailoring it to the agent's political literacy and disposition. The officer must be aware of how the agent relates to political matters and adjust accordingly.

Such conversations do not exhaust the full scope of ideological education. The case officer should also draw upon the agent's personal and professional interests, recommending books, films, plays, or exhibitions that align with socialist values. In all such cases, the principles of operational security and tradecraft must be rigorously observed. Officers should help the agent determine discreet ways to obtain and engage with materials, ensuring the agent does not draw undue attention or expose himself to counterintelligence interest through careless enthusiasm for such topics.

Ideological-political education of agents does not tolerate formulaic or mechanical approaches. Agents represent diverse social strata, political convictions, and intellectual horizons. Their motives for cooperating with the intelligence service of a socialist state also differ. Therefore, a differentiated approach is a mandatory condition for effective political instruction.

1. Education of Agents Recruited on an Ideological-Political Basis

An agent recruited by the intelligence service of a socialist state on ideological-political grounds is, to some degree, a conscious adherent of the socialist cause or a supporter of the political line pursued by socialist states. Such agents typically show interest in political issues, the domestic and international affairs of socialist countries, and achievements in science, technology, and economics. They are often intellectually engaged with Marxist-Leninist theory.

Such an agent views cooperation with intelligence as a form of political work. In these cases, the officer's task is to expand the agent's political horizons, teach him to properly assess international developments, and deepen his ideological convictions and commitment to the mission.

It would be a mistake to assume that this category of agents requires no further education. Even those in capitalist societies who support the socialist cause remain subject to the influence of their environment and bourgeois propaganda. They may harbor misconceptions, adopt misleading perspectives on political questions, or internalize bourgeois ideas about morality, nationalism, or patriotism.

Therefore, ongoing political instruction is imperative with this category of agents. The officer must remain alert and responsive to any manifestation of ideological inconsistency or confusion.

The officer must be capable of identifying any erroneous or conflicting ideological views held by the agent and of providing persuasive and accessible explanations. Special attention must be devoted to the agent's national sentiments. Under the influence of bourgeois nationalist and anti-communist propaganda, certain agents diminish the importance of class struggle, misinterpret the national interests of their own people, and misunderstand concepts such as morality and patriotism.

In some instances, such agents regard their cooperation with the intelligence service of a socialist state as espionage. In such cases, it is imperative to clarify the fundamental differences between the tasks and principles of socialist intelligence and those of capitalist services. The officer must explain that by cooperating with the intelligence services of a socialist country, the agent is participating in the defense of the interests of all working people, including those of his own nation and class.

A failure to properly convey this message, particularly regarding patriotism, can result in a loss of motivation or even a complete severance of contact. For instance, in one European country, a woman codenamed "Mira" was recruited by a progressive organization via a recruiter-agent. She was a local national working in a foreign embassy. A self-described patriot, she initially agreed to assist the organization in its fight against foreign interference in her country's internal affairs.

However, the recruiter assumed that Mira was sufficiently politically prepared and did not undertake any further ideological education. The case officer, relying on the recruiter's assurances, did not supervise Mira's development or monitor her operational reliability.

At first, Mira participated actively, assisting in several covert removals of classified materials from the ambassador's safe. Over the course of a year, she delivered valuable intelligence. Eventually, however, the case officer noticed a decline in the volume and regularity of her deliveries. Later, the recruiter reported that Mira had ceased cooperation entirely, stating that due to changes at the embassy, she no longer felt safe or trusted and could not continue working under those conditions.

A review revealed that Mira's withdrawal had a deeper

cause. Influenced by reactionary propaganda, she had come to believe that her assistance to the progressive organization no longer aligned with the interests of her country. As a result, she terminated the relationship.

In this instance, both the case officer and the recruiter had failed to provide Mira with proper ideological-political education. They did not explain that the long-term interests of her people were fully aligned with the goals of the working class and of all progressive humanity. This omission ultimately resulted in the loss of a valuable asset.

Another case involved an agent-handler who informed his case officer that a certain agent, codenamed "Dick", possessed a valuable invention related to long-range missile applications. He recommended that the officer meet directly with Dick to discuss the matter.

At the meeting, Dick claimed that there had been a misunderstanding and that the invention in question was still in its early developmental stages. His account contradicted that of the handler. The officer attempted to probe further, asking several technical questions related to the device. While Dick remained polite, he was markedly evasive throughout the exchange.

This aroused the officer's suspicions, and he again approached the handler. The latter was surprised by Dick's reluctance, since Dick himself had previously described the invention in detail and had even shared insights into its construction.

Further investigation confirmed that Dick had indeed completed the development of the device. It had passed field testing and had already been accepted into service. Dick, however, had chosen to conceal this fact from us.

This decision was influenced by conversations he had with the director of his firm and by the ideological environment of the workplace, which emphasized the national significance of the invention. During these conversations, and under the pressure of intensified propaganda against socialist states, Dick was persuaded to believe that sharing his invention with our intelligence service was inappropriate and that he should not share the invention. The KGB Station determined that Dick's reluctance stemmed primarily from his nationalist sentiments. Within his firm, the invention was regarded as a matter of national pride, and the idea of handing it over to a foreign intelligence service, even that of a socialist state,was equated with betrayal.

These nationalist convictions were not identified in a timely manner, neither by the recruiter nor by the KGB Station. Thus, when Dick was faced with a serious decision, he hesitated. Only through persistent political education was the officer eventually able to help Dick recognize his ideological error and reengage him in active cooperation with the intelligence service.

These examples demonstrate the critical importance of recognizing and addressing an agent's national sentiments. Timely intervention is necessary to prevent their drift toward narrow nationalism, and sustained effort is required to reorient them toward internationalist values. The goal is to instill in the agent a clear understanding of the connection between intelligence work and the fundamental interests of their own people, interests that are best served through the struggle of socialist states.

Experience shows that even ideologically close agents are susceptible to wavering or betrayal if they come under hostile influence and have not received sufficient ideological education. The following example confirms this point.

A husband and wife, codenamed "Charles" and "Linda", were recruited on an ideological-political basis and cooperated with the intelligence service for ten years. These were individuals ideologically close to us. Prior to recruitment, both had been members of a progressive organization, had considered joining the Communist Party, and were engaged in Marxist-Leninist study.

However, after their recruitment, despite recommendations from the KGB Station, they gradually withdrew from their former affiliations. They severed ties with their former comrades, ceased subscribing to progressive newspapers and journals, and stopped attending Marxist study groups. They moved to a different part of the city and began to adopt the lifestyle of the petty bourgeois.

They insisted was justified, as it was intended to cover their identities, expand their operational capacities, and prepare them for more sensitive assignments. However, it must be noted that progressive organizations in capitalist countries are often under active counterintelligence surveillance. As a result, any sustained agent contact with them becomes inherently risky.

During their ten-year cooperation with the KGB, Charles and Linda remained in contact with multiple case officers, most of whom neglected their ideological development, having prematurely labeled them as "model agents" due to their long-standing loyalty. However, their petty-bourgeois environment, the social setting into which they had fully integrated, began to exert a corrosive influence.

Though well-supported materially and operationally, and though their relationships with the intelligence officers remained formally intact, the agents' inner convictions began to erode. During this period, nationalist propaganda and the rhetoric of "spy mania" intensified in their host country. Bourgeois propaganda stirred patriotic sentiment among the population and equated cooperation with foreign services, even socialist ones, with treason.

Under such conditions, Charles and Linda underwent a political reorientation, ultimately culminating in outright betrayal.

When intelligence work requires an agent to operate in reactionary circles, gaining trust among politically hostile individuals, the agent, especially one without solid ideological grounding, may begin to conceal his prior views, avoid association with progressive individuals, and distance himself from socialist ideology. He may cease participation in progressive groups altogether.

In these circumstances, the agent is increasingly surrounded by reactionary influences, leaving him vulnerable to ideological drift and political contamination by the bourgeois press and propaganda. The intelligence officer is the one person with whom the agent can speak openly, seek clarification, and receive ideological guidance.

A significant portion of the agent network recruited on an ideological-political basis demonstrates not only interest in current political events but also in Marxist-Leninist theory. The intelligence officer must treat the agent's ideological inquiries attentively and with tact, leaving no question unanswered, and leveraging the agent's political interest to guide discussions towards relevant developments, theoretical doctrine, and expansion of the agent's political worldview. These efforts serve to reinforce the agent's ideological proximity to the socialist state.

The agent's political orientation must be continuously monitored through discussions on acute questions of international affairs. The intelligence officer should initiate such discussions when necessary but also respond promptly when the agent takes the initiative. Signs of political disengagement must be identified early and addressed systematically.

During periods of international tension, or following important policy decisions by the leadership of the socialist state, political-ideological work with the agent must be intensified. Only through continuous political-educational engagement can political backsliding and the risk of defection be prevented.

2. Political Education of Agents Recruited on Non-Ideological Grounds

Among agents recruited on a material basis, there may be individuals with a wide range of political views and sentiments. There are cases in which individuals sympathetic to the socialist state were recruited not on the strength of ideological affinity, but rather through material inducement. In such situations, the ideological factor served only as a secondary or auxiliary motive to secure the individual's cooperation with the intelligence service.

The task of the intelligence officer in conducting educational work with such an agent consists in reinforcing the agent's sympathy for the socialist country, broadening his intellectual horizon, and teaching him to evaluate political events through the lens of Marxist-Leninist principles. In this way, the ideological motivation behind his cooperation with the intelligence service is solidified.

It is a different matter if the recruitment, conducted on a material basis, involved a politically apathetic individual, one who harbors no inherent sympathy toward the socialist state, and who views cooperation with the intelligence services purely as a commercial transaction. In such instances, the task of ideological-political cultivation becomes significantly more difficult, since the objective is no longer to develop an existing political inclination, but rather to ignite such an interest from scratch, to foster emotional affinity toward the socialist cause.

A thorough study of the agent's worldview and personal attitudes is required. The officer must identify points of entry that might awaken an interest in political matters, encouraging the agent to reflect on political developments and gradually move closer to our ideological position.

One illustrative case of such ideological influence involved a politically indifferent agent. The agent, codenamed "Fuchs," was recruited solely on a material basis and took no interest in political affairs. He behaved like an average citizen, concerned only with his own well-being. "Fuchs" consistently maintained an attitude of political neutrality and dismissed the intelligence officer's attempts to engage him in political dialogue.

However, in speaking with the agent, the officer discovered that "Fuchs" had lost his eldest son during the Second World War. When the conversation turned to this son, the officer invited "Fuchs" to a discussion about the nature of war. It was revealed that although "Fuchs" harbored deeply negative feelings toward war, he considered it inevitable and, with visible anxiety, expressed the fear that he might lose his younger son in a future conflict.

The intelligence officer exploited this vulnerability as an opportunity for ideological-political influence on the agent. Cautiously and tactfully, without forcing his own views, the officer used concrete historical examples to expose the aggressive nature of imperialist policy, particularly that of the chief bastion of world reaction - U.S. imperialism.

At the same time, the officer explained the peaceful policy pursued by socialist countries, demonstrating the socialist state's sincere commitment to peace and its willingness to preserve it through collective efforts of all peace-loving nations, in contrast to imperialist forces that provoke war. He exposed for the agent the policy of appeasement practiced by the United States, Britain, and France toward fascist aggressors on the eve of the Second World War, highlighting the direct imperialist interest in global warfare.

As a result of this sustained and carefully calibrated educational effort, the agent began to approach political discussions more thoughtfully, initiated independent efforts to obtain classified materials related to his government's war preparations, and his operational activity became more purposeful and ideologically grounded.

In this way, the cooperation of an agent initially recruited on a purely material basis was strengthened through ideological alignment.

The intelligence services of socialist countries must not refrain from the ideological-political cultivation of agents who have been recruited on material grounds, even when their political convictions are hostile to progressive worldviews. Admittedly, this is a particularly difficult and delicate task. Where the target is an apolitical individual, the officer's objective is to awaken interest in political matters and to foster ideological commitment. In cases where the agent harbors views overtly hostile to the socialist state, the officer's task becomes that of actively replacing those views with sound political judgments and ideologically reorienting the agent toward socialist principles.

Naturally, while the methods and tools of this work remain largely consistent with those used elsewhere, operations in this realm demand significantly greater effort and attention on the part of the intelligence officer.

It is not uncommon for an agent's hostile attitude toward the socialist state to stem not from deeply held convictions, but rather as a consequence of exposure to bourgeois propaganda, ignorance of the truth about the socialist homeland, or simple misinformation. In such cases, explanatory work by the intelligence officer can yield positive results.

As with many other situations, the officer must speak to the agent truthfully and clearly, countering the distortions of bourgeois propaganda and presenting the principled position of communist and workers' parties on international matters.

In colonial and dependent countries, officers of the socialist intelligence services who work with agents recruited on material grounds must pay special attention to the agent's attitude toward imperialist colonizers. An agent who is inherently hostile to colonial powers will often exhibit a genuine interest in the national question and in the policies of socialist states related to national liberation. Such sentiments should be leveraged in the agent's political-educational upbringing.

More broadly, the task of the case officer is to conduct political and ideological work with agents recruited not only on material but also on moral-psychological grounds, including those recruited through the use of compromising materials. In these cases, especially during the early stages of post-recruitment interaction, relations are often founded on coercion, which may result in the agent feeling alienated, angry, or even hostile toward the intelligence officer.

Nonetheless, the intelligence service must not dismiss the potential value of such individuals. When circumstances allow for a realistic opportunity to transform the agent's mindset, educational efforts must be pursued with patience. Agents must be given time to adjust to the officer, to develop working habits, and to gradually engage in intelligence activities.

By establishing professional rapport, with attentiveness, respect, and human understanding, the officer should initiate conversations with the agent on topics that concern him personally or that involve people close to him. This creates the conditions for the agent to realize that he has no basis for animosity toward the socialist state. On the contrary, he comes to see that he is treated with consideration and care.

As the agent becomes more involved in operational activity and begins receiving material rewards or benefits, political engagement should intensify, and ideological reinforcement must follow in parallel.

The officer must present well-reasoned facts and examples that illustrate the inevitable collapse of capitalism and the historical necessity of its replacement by socialism. The officer should lead the agent toward understanding that, in choosing to work with the socialist intelligence service, he has taken the only correct path.

Haste and clumsy ideological coercion, including overt propaganda or poorly handled "re-education" efforts, can be counterproductive, as such methods are often perceived by the agent as dishonest or manipulative, especially when he is under the influence of hostile press and reactionary thought.

One illustrative example of ideological re-education involved a resident KGB station's efforts to rehabilitate an agent who had been recruited on moral-psychological grounds, using compromising material.

In a certain Western European country, the KGB station learned of a major in the American army, codenamed "Grey," who was undergoing severe financial difficulties. "Trent" served in a rear unit of a U.S. division stationed in the target country. Being in a senior staff position, he had access to sensitive documents, including materials related to missile deployment at a recently established American base.

After identifying "Grey" as a potential asset, the KGB Station obtained reliable information characterizing his personality and lifestyle. It was discovered, for instance, that he maintained an intimate relationship with the wife of a prominent pro-American political figure.

This affair was carefully concealed by "Grey," as he understood that if the extramarital connection with this government official's wife were to be exposed, even privately, he could not avoid serious repercussions from the U.S. State Department.

"Grey's" indulgences required heavy expenditures, significantly exceeding his official income. Having lost self-control, he began embezzling government funds, incurring a deficit which demanded substantial repayment. In his quest for money, he attempted to sell military materials on the black market, including sheet metal and components from the U.S. designated for missile site construction. However, no buyers were found for these goods. "Grey" was also described as a man with reactionary political convictions and was known to be emotionally unstable, frequently involved in scandalous episodes back in his home country.

Upon analyzing the dossier on "Grey," particularly the data regarding his embezzlement and the intimate affair with the government official's wife, the KGB Station decided to recruit him by using these materials. Public exposure of his affair would likely provoke severe repercussions from the U.S. State Department and possibly lead to his recall from the country. Since "Grey" had no means to repay the embezzled funds, and once confronted his guilt would be evident, his situation was further aggravated, especially given that criminal prosecution could ensue over the financial fraud.

Taking all this into account, it was decided that a KGB officer would meet with "Grey" and make a direct proposal for cooperation. For the scenario in which "Grey" might refuse, the officer was instructed to present him with the likely consequences: disclosure of several facts from his affair, compromising both his mistress and her husband. This would be immediately followed by evidence of his embezzlement and attempts to profit from black market sales of military materials.

The invitation to the meeting was delivered in a letter placed in "Grey's" mailbox. It was written in a courteous tone so that he would not suspect it originated from the intelligence service. The letter did not hint at recruitment, it mentioned the possible purchase of some materials from him, which would align with his known attempts to sell such items.

As the KGB Station had anticipated, "Grey" came to the meeting. Omitting the operational details of the recruitment conversation and the character of "Grey's" reaction, we focus on the outcome: "Grey" agreed to cooperate with the intelligence service under the condition that the agreement would remain valid only while he was stationed outside the United States. If he returned to the U.S., intelligence would be required to terminate contact and not pursue reestablishment of relations. The recruitment officer, not having fully resolved his operational task, was compelled to accept "Grey's" conditions, hoping that over time a more permanent arrangement could be established and these limitations overcome. Indeed, once "Grey" accepted the proposal, it was reasonable to expect that he might begin preparations to return to the United States, but this did not occur.

"Grey's" initial operational integration went smoothly. He delivered his first batch of valuable documents, for which he received a sum equal to approximately one-fifth the amount he was required to repay to cover his embezzlement. The purpose behind this relatively modest payment was to demonstrate to "Grey" the tangible benefits of cooperation with the intelligence service and to reduce his anxiety, which he understandably felt despite his otherwise decisive nature. Subsequent transfers of documents occurred in a calm atmosphere, as "Grey" began to acclimate to the intelligence relationship. This in turn allowed officers to begin his ideological-political indoctrination.

A KGB officer named Valentin was assigned to handle "Grey," with the ability to meet him in an official setting. Valentin had a solid understanding of the habits and lifestyle of American military officers, which enabled him to rapidly establish a working rapport with "Grey." However, Valentin's initial attempts to engage "Grey" politically yielded no positive results international affairs did not interest him, and he had no desire to discuss them. The only topic that did engage "Grey" was the question of commercial trade between American companies and the socialist country where the KGB officer operated. It soon became clear why: this topic was close to "Grey's" heart, as his father had been co-owner of a California company that, prior to the war, had exported rose oil to the socialist country.

Sensing "Grey's" interest in trade matters, Valentin gradually expanded the range of discussed topics and guided "Grey" toward the idea that commercial relations between nations cannot be separated from the broader context of intergovernmental relations. The natural development of this theme gave Valentin the opportunity to steer their conversations into political territory and to stimulate "Grey's" interest in those issues, even though his responses were, at first, indistinct and guarded.

Later, a revealing characteristic of "Grey" was discovered, which influenced his worldview. Under the influence of anti-communist propaganda, "Grey" had developed a firm belief in the inevitability of nuclear war, which, he believed, could erupt at any time. This led him to adopt a mindset that life must be lived only for today, without concern for the future, since it promised nothing good. In essence, "Grey" was morally disoriented and psychologically traumatized, a man who had lost a clear purpose in life. As became evident later, he had only a shallow understanding of political matters, but he avoided them deliberately, regarding politics as a shady business unworthy of a U.S. officer.

Valentin had to work patiently with "Grey" before he could become a reliable political ally. However, Valentin refrained from using compromising materials, as the relationship with the agent was developing smoothly and "Grey" was gradually becoming ideologically aligned and a dependable asset to Valentin.

3. Indoctrination of Agents of Influence

The primary task in handling this category of agents is their ideological and political education, which, under current conditions, assumes particular importance. This does not suggest that complete ideological alignment or doctrinal adherence to communist principles is a prerequisite for their operational utility. Rather, this category of agents is engaged by intelligence services to exert influence over the internal developments and foreign policy of the target country in directions favorable to the socialist system. Agents of influence are typically embedded in reputable professional, political, or public roles.

The interests of this category of agents, in many cases, partially align with the political objectives of the socialist state, for example, on issues such as the struggle for peace, nuclear disarmament, opposition to national oppression and racial discrimination, and resistance to the expansionist policies of imperialist powers. In working with such agents, the intelligence officer must strive to broaden the agent's outlook and to gradually achieve ideological alignment with our cause. If this goal is attained, the agent will become more proactive and productive in his activities.

The ideological indoctrination of agents of influence is also essential for ensuring the long-term reliability of their cooperation with intelligence and to prevent their potential defection to hostile camps.

In most cases, such agents are recruited from among prominent political, industrial, financial, public, and governmental figures, including emigrants and dissidents. Their cooperation with intelligence is often viewed not as traditional espionage, but as an independent political activity. An intelligence officer working with agents of influence must have a solid grasp of the internal and external politics of the target country, know its history, culture, customs, and legal system, possess sufficient life experience and operational acumen, and be capable of winning the trust of agents while applying the proper methods and forms of ideological-political education.

Using specific examples that are clear to the agent, the officer must convincingly demonstrate the historical inevitability of capitalism's collapse and the advantages of the state and social systems of socialist countries. He should explain the peace-oriented policies of socialist states and expose the aggressive nature of the policies of imperialist powers. Through skillful comparisons of the two systems, the intelligence service can exert meaningful political influence on the agent and promote his ideological development in the spirit of loyalty to the cause of the socialist camp.

When conducting educational work with this category of agents, the officer must act with particular tact, bearing in mind that he is dealing with individuals of high political and social status. Overemphasis on the "importance" of their role may alienate them or cause resentment, especially if they perceive themselves to be grouped with lesser agents. This may even lead to refusal to cooperate. At the same time, the officer must avoid encouraging vanity or the belief that the agent is irreplaceable to the intelligence service of the socialist state, this could negatively affect operational reliability.

The goal of such work is to overcome the agents' temporary personal biases and narrow interests and to broaden their capacity to serve tasks aligned with advancing the objectives of the socialist state. This includes influencing selected elite circles, individuals, and aspects of life within capitalist countries in ways favorable to socialist goals.

The correct selection of forms and means of political persuasion, along with perseverance in achieving objectives, significantly determines the intelligence officer's success in the ideological and political indoctrination of the agent.

Let us illustrate this with a concrete example of an intelligence officer's work in politically reeducating an agent of this category.

During the Second World War, a prominent public figure from one of the Western European countries, hereafter referred to as "Gustav", was living in exile in Sweden. A vocal anti-fascist, Gustav wrote articles and gave radio speeches against Hitler's Germany and against National Socialism.

"Gustav" came from a well-known bourgeois family, was a "respectable" Catholic, and had previously held a significant position within the Christian Democratic party in his homeland. It was reasonable to assume that after the war, with the liberation of his country, Gustav would once again become an influential political figure. Based on this, the intelligence service decided to recruit him.

The officer designated to handle the recruitment was an operative named Aleksandr, who had maintained official contact with "Gustav." Upon Aleksandr's proposal for covert cooperation, Gustav immediately stated that he was willing to assist in gathering information on political matters, but asked not to be considered an intelligence agent.

Recognizing that "Gustav" was a committed anti-fascist and a progressive political figure, Aleksandr deemed it essential to begin steady ideological indoctrination of the agent and to limit interaction strictly to the framework of Gustav's practical anti-fascist activities. Over time, between "Gustav" and Aleksandr a relationship developed that, while not agent-based, became genuinely trusting. As such, the relationship was documented as an unofficial operational contact.

After the war, "Gustav" returned to his homeland and was contacted by KGB Station officer Konstantin. At that time, parliamentary elections had been held, and a Christian Democratic government had come to power. From the very beginning, the new government pursued a pro-American course.

Taking into account "Gustav's" influence within liberal-bourgeois circles, the progressive intelligentsia, and among left-leaning Catholics, who were in opposition to the government's foreign policy direction, the intelligence service intended to use him to establish a new liberal-bourgeois party with a progressive orientation.

Shortly thereafter, "Gustav" received a proposal from the leader of the Christian Democratic Party to rejoin its ranks and publicly endorse the government's political course. "Gustav" was promised a high-ranking position within the state apparatus. He relayed this offer to Konstantin, stating that he found it acceptable. When Konstantin inquired whether "Gustav" shared the Christian Democrats' political platform and the government's pro-American stance, "Gustav" replied that, although he did not yet fully understand the party's political trajectory, it could not be entirely regressive in his view. After all, the party had been banned under the Nazis, and many of its leaders had suffered repression. Moreover, he felt he could not oppose the government's foreign policy orientation, given that his country had been liberated from fascism by the Anglo-American armies.

It became evident that "Gustav" was on the wrong path, misjudging the political situation in his country and misunderstanding the roles of the U.S. and Britain both during and after the war. He was in clear need of intensive political reeducation. Under these circumstances, the immediate task for the intelligence officer was to prevent "Gustav" from accepting the

Christian Democratic leader's proposal. Direct prohibition was not an option, it would likely have produced the opposite effect, especially since "Gustav" did not consider himself an intelligence agent, and his relationship with the service was still informal.

Konstantin tactfully advised "Gustav" to refrain from accepting the offer extended to him, justifying his position by stating that "Gustav" had not yet familiarized himself with the domestic political situation in the country and did not fully understand the political course of the Christian Democratic Party. Furthermore, Konstantin expressed his concern that the party's leadership had seen fit to make a public announcement in the press concerning "Gustav's" support for its political program.

"It seems," Konstantin said, "the party intends, on one hand, to associate itself with your authority, and on the other, to exploit your statement to bolster its image, an image that, as you know, has weakened recently due to the government's failure to address the country's deepening economic crisis."

"Gustav" took time to reflect on Konstantin's arguments and, at the end of their discussion, promised to postpone any decision regarding the Christian Democrats' offer and to study the political climate of the country more thoroughly. As an interim step, Konstantin recommended that he pay attention to the trial of the known collaborationist T., particularly the role played by American representatives in shielding him.

At their next meeting, "Gustav" arrived in a very agitated state. He angrily recounted that, according to a lawyer acquaintance, the government was eager to suppress the trial of T. because it possessed compromising information about the wartime conduct of several members of the ruling establishment during the occupation. These individuals could be exposed if the case proceeded. Konstantin confirmed "Gustav's" information and clarified that not only the national government, but also American interests were involved in the cover-up. The Americans, he explained, were protecting one of the Christian Democratic Party's top leaders, a minister of internal affairs, who was the subject of compromising materials and whose continued position was considered strategically important by the U.S.

In the course of subsequent meetings, Konstantin, relying on specific examples, led "Gustav" to the conclusion that the leadership of the Christian Democratic Party was composed of elements far removed from progressive ideals. It was therefore essential to uncover their true agenda and behind-the-scenes activities. "Gustav" fully agreed with Konstantin and explained how and through whom he could obtain information regarding the internal state of the party and the secret plans of its leadership.

Thus, Konstantin achieved two critical objectives: he involved "Gustav" in active intelligence work, which was valuable not only for acquiring information but also for securing "Gustav" as an agent, and simultaneously enabled him to personally observe the reactionary nature of the Christian Democratic Party's political course.

Through his longstanding connections with influential party members, "Gustav" was able to obtain insider information on the party's internal dynamics, its future plans, and covert links between its leadership and American officials. These revelations increasingly convinced him of the Christian Democratic Party's reactionary character and its collaboration with U.S. political interests. Consequently, he rejected the offer from the party's chairman.

To further expose the Anglo-American duplicity during the Second World War, Konstantin advised "Gustav" to read a book written by a former British War Office official titled How We Planned the Second Front. This publication revealed how British and American authorities had, in fact, delayed the opening of the second front. Konstantin's calculation proved correct: the book deeply affected "Gustav," especially as his own brother had died in late 1944. "Gustav" believed that a timely Allied offensive could have saved his brother's life, thereby intensifying his personal disillusionment.

As Konstantin continued analyzing the economic and political measures carried out in the country by the Americans through the local government, he led "Gustav" to the conclusion that the Americans aimed to subjugate his country economically and politically, depriving it of its national sovereignty. At the same time, Konstantin used every opportunity to engage "Gustav" in discussions about the peace-oriented foreign policy of socialist states and the fraudulent nature of bourgeois democracy. Konstantin's high level of political preparation enabled him to emerge victorious from the ideological debates he engaged in with "Gustav."

As a result of Konstantin's carefully considered and methodical educational effort, he succeeded in transforming "Gustav" into a convinced ideological supporter of the socialist state and a staunch opponent of Anglo-American imperialism and the reactionary government of his own country. This subsequently allowed the KGB station to use "Gustav" for the formation of a liberal-bourgeois party that advocated national independence and the establishment of friendly relations with socialist countries.

4. Ideological Indoctrination of Group Leaders and Recruitment Agents

Group leaders are agents responsible for managing the intelligence activity of agents under their control. Simultaneously, they must carry out their ideological-political upbringing. Therefore, in both political and operational matters, they must maintain authority over their subordinate agents.

The intelligence services of socialist states possess many examples wherein politically mature group leaders not only effectively fulfilled reconnaissance tasks but also successfully managed valuable agent networks.

Intelligence officers must not only indoctrinate group leaders but also encourage them to carry out similar educational efforts with their own agents. In some cases, it is necessary to ensure that a group leader, who may have agents under his control and who himself may originate from a hostile social class, becomes a conscious collaborator with the intelligence services of the socialist state and dutifully fulfills assigned tasks. Such results can only be achieved by a group leader (agent-gruppovod) with a broad political outlook, one who correctly understands both the foreign and domestic policies of the socialist states.

Group leaders and recruiting agents in most cases lack experience in ideological-political education. Therefore, intelligence officers must not rely solely on them for the political development of each agent in their charge.

When it comes to indoctrinating agent networks run by group leaders, intelligence officers must not depend solely on the latter. They must periodically meet personally with such agents and use these meetings as opportunities for ideological-political instruction. The practical assistance rendered to the group leader during this process facilitates the ideological education of the agent, increasing her efficiency, reliability, and loyalty.

5. Particular Features of Ideological-Political Indoctrination of Agents Recruited from Among the Emigré Communities

Ideological-political indoctrination is of particular importance when it comes to agents recruited from among émigré communities. It is essential to keep in mind that such agents have, for the most part, endured significant hardships during their emigration and have personally experienced all the "delights" of life in capitalist countries. Nevertheless, one must always consider that some agents, while residing in capitalist states, have become infected with the vices of bourgeois society, corruption, sycophancy, careerism, moral unreliability, materialism, etc.

Moreover, this category of agents is often under strong influence from various forms of bourgeois propaganda and remains in constant contact with declared enemies of socialist countries. If proper ideological instruction is not provided, such agents may revert to the enemy camp.

It should also be taken into account that émigrés' attitudes toward socialist countries fall into three principal categories:

The first group of émigrés does not belong either to reactionary or progressive organizations and consists partly of the general population of capitalist countries.

The second group comprises persons with clearly progressive views.

The third group consists of antisocialist elements affiliated with reactionary émigré organizations hostile to socialist countries.

When working with émigré agents, it is essential first and foremost to cultivate in them patriotic sentiments, instill a sense of human dignity, and rekindle love for the Motherland. Agents must be told about the achievements of the socialist state in the fields of economics, science and technology, culture and the arts, and about the changing living conditions of the people in that country. Particular emphasis must be placed on regions and localities where the agents were born and lived prior to emigration. Agents must be shown, using facts, the growth in popular well-being and be made to compare this with the actual living conditions of the majority population in the countries where they now reside.

If constitutional circumstances permit, agents should be encouraged to read, at least occasionally, literature that explains these issues. In some cases, letters from the agents' relatives residing in socialist countries may also be used for educational purposes. Such letters inform agents of the positive changes taking place in their homeland under socialism.

In the process of political-educational work, it is also important to explain to these agents the relevant legal decrees issued by the socialist authorities concerning amnesty for emigrants and displaced persons. Such conversations may awaken in the agents a sense of duty to the Motherland and a realization of the need, and the possibility, to atone for their guilt and become worthy citizens of the socialist fatherland.

Properly conducted work with such agents leads to them becoming reliable and loyal members of the agent network, carrying out assigned tasks with greater willingness.

When engaging in ideological-political education of agents recruited from hostile émigré communities, it is always necessary to remember that the local counterintelligence services monitor such émigré organizations and, for preventive purposes, periodically place certain individuals under surveillance. Surveillance of individual émigrés may even be carried out under the direction of the émigré organizations themselves, if they suspect those individuals of involvement in activities directed against the organization. From this it follows that agents recruited from émigré circles must be thoroughly instructed on how to conduct themselves.

The intelligence services of capitalist countries are highly active among émigré populations, seeking to identify individuals suitable for subversive work against the socialist bloc.

All these factors must be carefully considered when working with émigré agents. Special attention must be paid to the security of both the intelligence officer and the agent.

6. Specific Aspects of Ideological-Political Indoctrination of Agents Handled by Illegal Station Officers

The principles of ideological-political indoctrination of this category of agents differ little from the principles used for agents managed by "legal" KGB stations. However, a key advantage of an illegal officer's relationship with the agent lies in the greater freedom to conduct face-to-face meetings. This enables the officer to better understand the agent and to influence him in the desired direction for the operation.

The situation is quite different, however, when the illegal officer or group leader is working with an agent who was recruited under false pretenses. For instance, if an agent has been recruited under the guise of working for some capitalist intelligence service, the officer's task becomes first and foremost to prevent that individual from becoming an enemy of the socialist state. The illegal officer must conduct a comprehensive study of such an agentt, to clarify his convictions, moods, and views, in order to determine how best and most effectively to guide the agent toward the desired intelligence and political influence objectives. In conversations concerning political matters, the illegal officer should maintain an outwardly neutral position, refraining from imposing personal views upon the agent, and instead using carefully selected factual material to lead the agent logically to the correct conclusions. The illegal officer must find common ground with the agent on a particular issue, then deepen and expand upon that shared viewpoint.

Political indoctrination of agents recruited under a false flag requires the illegal officer to display exceptional skill and inventiveness. The difficulty lies in the fact that the agent must never suspect he is working for the intelligence service of a socialist state. Such work must be conducted prudently, flexibly, and discreetly, with constant emphasis on the paramount need to ensure the security of the officer himself and, by extension, of the entire intelligence organization.

To avoid operational compromise, the officer must never permit rash or unconsidered actions. For example, if the agent recruited under a false flag expresses neutral or indifferent attitudes toward social systems in general, the officer should, through subtle persistence and flexibility, gradually cultivate a sympathy toward socialism and ultimately turn the agent into a political ally.

Particular flexibility and careful attention to context are needed when exerting political influence on agents with reactionary leanings. For instance, they might be encouraged to read Marxist-Leninist literature under the pretense that "it is necessary to know one's enemy." By playing the role of a person with "independent and broad views," the officer can instill in the agent a skeptical attitude toward his government, gradually leading him to critically examine the world around him, the surrounding reality, and then gradually lead him toward a critique of bourgeois values, from a Marxist-Leninist position, such that the agent gets the impression that this critique arises organically from his own thoughts and worldview.

Attempts to politically re-educate agents who are avowed enemies of socialism may not be appropriate at all; in many cases, such efforts can provoke resistance and lead to operational failure with serious consequences.

The political development of the agent is the central element of the officer's work in ideological-political education. However, alongside this, the officer must exert influence on the moral character of the agent, cultivating in him such qualities as honesty, truthfulness, principled behavior, initiative, and courage. In this regard, the officer's personal example assumes paramount importance. While fostering courage in the agent, the officer must also be scrupulously attentive to his own conduct during meetings. If the officer appears nervous or constantly looks around during contact, this will immediately transmit to the agent, instilling in him a sense of insecurity. Consequently, the officer's words about courage, resilience, and composure will have no meaningful impact on the agent.

PART III

TRAINING AND GUIDANCE OF AGENT OPERATIONS

Introduction

The instruction of an agent in the methods of conducting intelligence activities and their ideological development in the spirit of loyalty to the intelligence service is a unified process.

Agents of foreign intelligence who are entering into covert cooperation for the first time typically lack knowledge of tradecraft and are unfamiliar with the techniques of conducting intelligence work. Therefore, the operative must train the agent in clandestine procedures, teach him how to execute assigned tasks, instill discipline, planning skills, and precision, and cultivate initiative in his operational work. The goal is to enable the most effective exploitation of the agent's intelligence-gathering potential.

Training the agent in tradecraft is one of the most critical prerequisites for operational cooperation. The success of any intelligence engagement depends above all on the extent to which both the case officer and the agent observe the full spectrum of tradecraft protocols. The degree to which the agent understands and applies clandestine operational techniques, and his ability to conceal his connection with the intelligence service, directly affects not only the agent's personal security but also, in many cases, the security of specific nodes within the broader intelligence network.

If the case officer establishes contact with an experienced agent who has collaborated with the intelligence service of a socialist state for a long period and already possesses specific skills in tradecraft, then it is, naturally, unnecessary to begin his training from scratch. In such cases, the operative must verify the degree to which the agent has mastered clandestine techniques, fill any existing gaps in knowledge, and then continually assess whether the agent is applying those techniques in practice.

When dealing with a newly recruited agent, the case officer must assume he is working with someone who is entirely inexperienced in the matters of tradecraft and, due to this, may commit reckless acts. This compels the officer to provide the agent with thorough instruction in the basic rules of clandestine activity, since ignorance or noncompliance with these principles may result in severe consequences for both the agent and the intelligence service.

The agent must be taught covert behavior, especially how to conceal intelligence activity and to keep his connection with the intelligence service secret from both colleagues and family. The officer's first goal must be to ensure the agent understands the absolute necessity of maintaining strict secrecy, even of the mere fact of their connection to intelligence.

This must be conveyed to the agent in the most concrete terms: even revealing his connection to the intelligence service to his closest friends or relatives, including his wife, could result in the agent being exposed by local counterintelligence services or law enforcement.

It should be understood that the agent, operating in the conditions of a capitalist state, must fully grasp that if his connection to a socialist state's intelligence service is uncovered, whether by the local authorities, counterintelligence agencies, or hostile political circles, he faces real danger. Lacking experience and essential skills, agents often fall victim to their own lack of operational caution.

Establishing contact with an intelligence service is a significant event in a person's life. Due to natural inclinations toward communication and openness, an agent may hasten to share this information with those closest to him. If the recruitment took place on ideological-political grounds, this confidant may be a like-minded individual, a wife, or another person the agent trusts. However, if the agent has friends and associates to whom he confides, and those people in turn have their own circles with whom they share news, the agent's secret can become known to several individuals and potentially come to the attention of counterintelligence.

The following example clearly illustrates this point:

A young woman arrived at the safe house of an intelligence service in France belonging to a socialist country. She had obeyed the protocols governing agent summonses and, in an anxious voice, urgently requested a meeting with the intelligence officer, stating that her matter was of the utmost importance and could not wait.

Upon arriving at the safe house, the case officer discovered that the woman present in the room was a complete stranger. However, noting that she had correctly given the password, the officer decided to proceed with the meeting. The woman immediately began with accusations: "You dragged my husband into a spy operation, and now he faces prison, or even death!"

Disturbed by the sudden turn of events, the officer asked who she was and why she believed she was connected to the operation. The woman burst into tears and said:

"Look, I know everything. I am the wife of 'X,' whom you recruited two months ago. Yesterday, my husband was arrested. He's an honest man, a good family father, and I am certain he committed no crime. But he's been detained, and they say it's for espionage. I beg you, don't pretend you don't know anything, please help me save Willy!"

The intelligence officer, after posing several questions to the woman, became convinced that she was indeed the recently recruited wife of agent "X," a man with significant and valuable intelligence potential.

The officer cautiously began to ascertain the full circumstances of her husband's arrest. In response to his questions, she stated that already on the second day after her husband's recruitment, she noticed a change in his demeanor and managed to get him to confess everything. Defending her husband, the woman explained: they loved each other so deeply that they were incapable of hiding anything from one another.

It turned out that the agent had informed his wife not only about his cooperation with the intelligence service, but also about the assignments he was fulfilling. Specifically, it was revealed that prior to his arrest, he had copied an important document and passed it to his handler. His wife had typed it on his typewriter. She also disclosed the address of the safe house and the contact procedure for arranging meetings with the case officer.

When the officer suggested that "X" may have told someone else about his involvement in espionage, the woman firmly denied it. She insisted her husband was naturally reserved and that it had taken great effort even for her to get him to divulge anything.

Aware of how the meetings with "X" had been conducted under strict tradecraft protocols, the officer was confident that the arrest could not have resulted from operational exposure. He thus began to press the woman for names of individuals to whom she might have spoken. Eventually, she admitted that she had casually mentioned her husband's secret to an old acquaintance, a seamstress, while being fitted for a dress.

After this admission, it became clear that the cause of agent "X"'s exposure was the woman herself. Once all the details of her conversation with the seamstress were known, the intelligence officer made the decision to instruct her on her future conduct.

He explained that only she was to blame for her husband's arrest and that now she must do everything possible to save him. The officer advised that, should she be summoned for interrogation, she should claim that she had merely repeated to the seamstress a fictional version she had read rom a detective novel, and that she had done so with the sole purpose of convincing her seamstress friend of her credibility.

The case officer was forced to abandon the use of the safe house and to categorically prohibit the woman from ever appearing there again. The "X" affair concluded successfully: he was released, as the accusation could not be proven; however, the intelligence service lost the ability to obtain valuable information from him.

Training an agent in tradecraft and intelligence procedures

demands that the officer strictly adhere to the principle: the agent should only be instructed in those methods and forms of intelligence activity that are strictly necessary for completing their assigned operational tasks. If the agent is not being groomed for recruitment activities, then he should not be instructed in recruitment procedures, techniques, or methodology. There is no need to teach the agent tradecraft communication protocols that he will not be expected to use in practical work.

At the same time, the agent must be thoroughly trained in all methods of intelligence work that are required to fulfill the intelligence tasking. In the course of working with an agent, it is essential to teach him how to properly write and format his reports, and to recommend that he prepare them immediately before meetings, either prior to a dead ketter box or right before meeting his handler.

As for the technical aspects of report preparation, the agent must be instructed to use ordinary paper, readily available at most retail outlets, and standard black ribbon or carbon paper when using a typewriter. Reports should never be written on personalized stationery or on the kind of paper the agent uses for personal correspondence. Nor should he use thick, high-quality paper, as this is more difficult to fold and takes up excessive space.

Equally important is that the formatting of the report correspond to an innocuous form, one that would provide no immediate basis for identifying the writer as an agent or linking him to espionage activity should it fall into the hands of counterintelligence. Such a form might be a "report" prepared by the agent under the guise of a casual line, such as "correspondence" to a newspaper, a "letter" to a friend, relative, and so on. In such documents, the agent may omit particularly sensitive moments, facts, names, details, etc. These can be recorded separately on a small slip of paper (which can be easily destroyed), or conveyed orally at the next meeting.

The agent must also be taught how to skillfully hide reports prepared at home or in a room, and be given advice on how to construct a home-based hiding place, essentially a materials cache. When instructing the agent on what to carry and how to transport it to the meeting or the dead letter box, it is advisable to recommend the use of disposable containers that can be easily discarded in the event of danger, thereby avoiding arousing suspicion.

An agent who is capable of delivering original documents for intelligence evaluation or copying, and which possess significant operational interest, must be taught how to handle such materials. The objective is to ensure the agent can distinguish between important and secondary documents and deliver the intelligence with minimal risk.

For example, if an agent is instructed to extract a certain classified document from a safe for handover, it is necessary to discuss in advance all potential methods for carrying out the task. The agent must be instructed that a document removed from a foreign safe or desk should be returned to the same exact spot from which it was taken, as quickly as possible. This means the agent must first carefully memorize the document's position. The document should not be crumpled, soiled, or creased, especially if it previously bore no such markings.

It is recommended that no staples or paperclips be removed from the document, and under no circumstances should it be altered. If the agent did not previously have authorized access to the document, i.e., had not handled it before, it is advised that he wear gloves when touching it, to avoid leaving fingerprints.

Particular caution must be exercised when handling cipher materials, especially cipher keys. It is known that cipher officers often leave a control slip, typically a piece of photosensitive paper, between the pages of cipher material stored in safes. When exposed to light, this paper changes color, thus indicating tampering.

Smoking while working with documents is strictly prohibited, as even a stray spark from a cigarette can irrevocably destroy critical material. In addition, one must never leave inkwells, bottles of ink, or any other liquids on the table where photography or other document handling is taking place. There have been instances where an intelligence officer, absorbed in his work, accidentally spilled ink and stained the document.

Agents tasked with copying sensitive documents must be trained to do so with great attention to detail and not to overlook any resolutions, marginalia, or other marks on the document. If appropriate, the agent should be taught to use a camera to record intelligence materials. If an agent does not already possess a camera, one should be issued or funds provided for him to acquire it himself.

The type of camera and the methods for concealment, including the use of microphotography and soft film, must be tailored to the specific operational environment in which the agent is working. In cases where an agent, for whatever reason, cannot have a camera, it is necessary to photograph the document via contact method directly on a photosensitive surface placed on a table. (Ideally, this should be a technical film with high contrast.)

Photographing documents has become firmly entrenched in the practical tradecraft of intelligence work and, in most cases, has proven extremely effective. If a document is photographed, there is no need for the agent to return it, thus eliminating the need for a second meeting and the risks associated with re-delivery.

An equally effective method is to have the agent dictate the document's contents onto magnetic tape and deliver the recording.

The agent must know how to remove and then restore documents in a safe if they were originally present there. It is essential to always coordinate in advance when the agent will remove the document, when and where he will hand it over to the intelligence officer, and how and when he will return it to the safe or desk. Finally, it must be determined how the agent will inform the officer that the operation has concluded successfully.

If a classified document is found in the possession of the agent, he must have a plausible cover story. Consequently, agents must be trained to carry out intelligence tasks with proper concealment measures in mind.

Within the limits of operational necessity, the agent should be made familiar with the methods of counterintelligence and police agencies. This is particularly important because many agents believe in the formal inviolability of such constitutional rights as the privacy of correspondence, confidentiality of bank deposits, telephone calls, and so forth.

The officer must train the agent in methods of detecting surveillance and techniques for evading it when necessary. Some agents mistakenly believe they can identify surveillance without training, and their behavior tends to draw unnecessary attention from others.

Sometimes the opposite problem arises: an agent, especially one recently recruited, assumes he is under constant surveillance. Such agents may panic and make critical errors, for example, discarding documents in their possession. It requires patient instruction to ensure the agent masters the skills of detecting and evading surveillance. Once these skills are instilled, the officer can conduct meetings with greater confidence in the agent's personal security.

It should also be noted that in recent years, counterintelligence services in several capitalist countries have intensified their operations against the socialist states and sometimes circumstances arise under which the intelligence officer, operating from the position of a "legal" KGB Station, is forced to miss meetings due to intensive surveillance conducted against him. In order to avoid such missed encounters, some KGB stations resort to establishing parallel contact arrangements with the agent, assigning two officers to work with him. If the first officer falls under surveillance and cannot meet at the appointed time, the second officer takes his place. This second officer, like the first, is fully briefed on the meeting and all related matters.

To create favorable conditions for the officer heading to a meeting, the Station may send several additional operatives into the city to disperse or mislead the surveillance and give the officer a chance to travel unnoticed to the designated area. It is also possible to draw the adversary's surveillance services away by deploying known faces connected to prior intelligence operations, provided the Station has such personnel at its disposal.

However, before transferring an agent to a parallel contact

setup, it is necessary to explain the rationale to him. This tactic, employed by KGB stations, is fully justified, as it minimizes or prevents missed meetings altogether.

Introducing agents to the working methods of hostile counterintelligence services must be done carefully and skillfully. It is vital not to confuse or alarm the agent, nor to lead him to think that failure is inevitable, which could drive him to withdraw from cooperation.

The officer must independently assess the extent to which the agent can be familiarized with counterintelligence methods. This largely depends on the agent's reliability, personal qualities, and the duration of his cooperation. When dealing with an agent who is still new to intelligence work and displays understandable caution, it is inappropriate to immediately reveal all the techniques employed by enemy counterintelligence that may only be necessary for seasoned operatives.

Constant vigilance, adherence to all requirements of tradecraft, caution, and understanding the methods of local counterintelligence significantly reduce the chances of either the officer or agent being compromised. However, compromise can still occur, and the intelligence officer must regard this as a real possibility, whether it results from mistakes made by the agent or from infiltration of moles into the agent network. Additionally, an agent may be detained or arrested for coincidental reasons. Such incidents may stem from preventive actions by local authorities during periods of internal political unrest in the country.

Given these risks, preparing the agent for the possibility of detention or arrest becomes vitally important. Operational experience demonstrates that a lack of proper preparation in this regard can lead to serious consequences. An agent might commit grave errors both during arrest and under subsequent interrogation, greatly increasing the chances of exposure. It is therefore imperative that the intelligence officer thoroughly instruct the agent in anticipation of possible detention or arrest.

However, such preparation must be approached with utmost caution. Clumsy or ill-timed instruction may leave the agent with the impression that arrest and repression are inevitable. For this reason, before discussing arrest contingencies, the officer must carefully assess the agent's mental state, find an appropriate context to broach the topic, and choose a suitable time to conduct the conversation.

If the agent has not previously been engaged in discussions of the dangers of clandestine activity, one must avoid directly implying the possibility of arrest unless absolutely necessary. If the agent is already sufficiently involved in intelligence operations, the officer must raise the subject with extreme care and in a delicate form, preparing the agent to accept that he may be arrested or detained even without serious cause, simply due to chance or circumstantial evidence. His understanding of this, and especially his conduct in such situations, will critically influence the final outcome.

The intelligence officer must further acquaint the agent with the investigative methods used by counterintelligence and police agencies during interrogation, and warn him about various provocations that may be employed by investigators and counterintelligence officers. These include: the use of a cellmate mole, face-to-face confrontations, cross-examination, covert surveillance within the detention facility, the planting of documents and evidence, and demands for a confession without formal charges being presented.

The agent must be instilled with the understanding that resolve and firmness during interrogation will positively influence the outcome of the case. He must not succumb to provocation or deviate from the pre-established cover story. Any inconsistencies in his testimony will only complicate his situation and may lead to operational exposure.

The agent must be taught how to securely store and conceal materials that are to be handed over to the intelligence officer, ensuring they are not discovered during a search. If the materials cannot be disguised due to their nature, he must be instructed in methods for disposing of them immediately in the event of imminent arrest or danger, whether by destruction, discarding, ingestion, or exposing film to light.

In certain cases where the agent, due to the nature of his official position, legitimately possesses documents that are of interest to intelligence, it is usually wiser neither to conceal nor attempt to dispose of them in the event of arrest. Instead, the agent should declare that he had brought them home in preparation for routine work, even if his access to such documents outside the workplace is prohibited by official regulations.

Instruction on agent behavior in the event of arrest or detention must also include ideological preparation to instill courage and the assurance that the intelligence service will not abandon the agent or his family in times of need. The agent must be convinced that the organization, being sufficiently strong, will be able to provide the necessary assistance to both him and his loved ones.

When training the agent in tradecraft and the methods of conducting intelligence activities, the officer must also bear in mind that certain agents, especially those new to intelligence work, tend to underestimate the dangers inherent in conducting intelligence operations and adhering to proper tradecraft. They may find the rules imparted by the case officer unnecessary or contrived. It requires considerable patience and effort on the part of the officer to instill in the agent the essential tradecraft habits and to ensure the agent understands that strict adherence to these rules is directly tied to both operational success and personal safety.

1. Cover Identity and Concealment of the Agent's Intelligence Activities

Proper establishment of a cover identity for intelligence activities and the concealment of the agent's operational tasks are of paramount importance for both security and mission success. This must always be a primary concern for the case officer.

To be effective, the agent's cover story must be plausible and must not arouse suspicion among those who may observe or inquire into his behavior. A good cover story must usually include a substantial element of truth. For this reason, it is important to be able to substantiate the cover story with actual actions.

When the agent's cover includes an acquaintance with the intelligence officer, it is sometimes necessary to stage an encounter between them at a reception, a public lecture, a conference, a fishing trip, a hunt, or similar activity. The establishment of a credible acquaintance is especially critical if the agent may, in the future, need to explain to someone his meeting with the intelligence officer. In support of his explanation, the agent can then cite a real-world event where they met, thus grounding the legend in actual circumstances.

When preparing to meet the intelligence officer, the agent must mask his departure from home by associating it with a believable activity, especially in the eyes of his spouse or relatives. Examples include going to the cinema, a club, a bathhouse, visiting friends, dropping by a newspaper office, library, or doctor. The agent should follow through with this activity either before or after the meeting in order to make his legend believable. Visiting the legend cover location reinforces its credibility, making the explanation plausible should it be questioned.

The same principles apply to the cover story for agent travel to another city or to the countryside, if a meeting with the intelligence officer is scheduled in that location. In all such cases, the agent must have a plausible and natural justification for the trip.

It is sometimes necessary to develop a separate legend for each meeting between the agent and the intelligence officer, depending on the location, timing, and the agent's public or professional circumstances.

The following example demonstrates what can happen in the absence of a prepared cover story for such a meeting:

An intelligence officer named Taras, stationed abroad at the diplomatic mission of a socialist country, was meeting with an agent known by the alias "Konrad" in a restaurant. Unexpectedly, the police entered the establishment, cordoned off the area, and requested identification from all patrons. Upon checking documents, a police officer became interested in the joint presence of the diplomat from the socialist country and a local resident who was known to be employed by one of the domestic ministries. Each man was questioned separately about whether they knew each other and why they were together in the restaurant.

Since no prior legend had been worked out, the intelligence officer and the agent gave contradictory accounts. The officer, mindful that he had no legitimate grounds for acquaintance with "Konrad," stated that he did not know him and had ended up at the same table by coincidence. "Konrad," on the other hand, claimed he was acquainted with Taras and had been invited to lunch by him. Naturally, such conflicting statements aroused police suspicion, and "Konrad" was detained.

Unprepared for interrogation, the agent gave confused and inconsistent testimony under questioning. Eventually, he became flustered, responded poorly to provocative questions, and ultimately confessed that he had maintained a clandestine relationship with Taras.

It must be kept in mind that if the intelligence officer holds diplomatic status, then in the event of his detention during a meeting with the agent, he may refuse to provide any explanations. This, in turn, gives the agent the opportunity to independently formulate a more advantageous account of the reason for the meeting or acquaintance with the officer.

If the meeting with the agent takes place at a safehouse (either a separate house or an apartment within a residential building), then both the agent and the officer must be prepared with a cover story explaining the visit. Naturally, the agent must know who the host is, if such a person exists. Depending on the host's occupation, an appropriate legend about the acquaintance and purpose of the visit can always be developed.

Maintaining tradecraft requires careful handling of the funds received by the agent from the intelligence service. In the event of an investigation arising from suspicion, counterintelligence typically examines whether the individual's spending aligns with their official income. Any discrepancy in expenditures is regarded by counterintelligence as a suspicious circumstance potentially linked to espionage or criminal activity. The agent must be trained to mask the use of funds received from the intelligence service and to convincingly explain the origin of any additional financial resources. This is a critical and responsible component of agent training.

Agents, like all people, have desires, needs, tastes, inclinations, and habits. Some save money to buy a car, a house, an apartment, a piano, or a camera. Others enjoy dressing fashionably or hosting acquaintances for lunch or dinner at an elegantly set table. The intelligence officer must, together with the agent, develop cover stories and plan activities that credibly explain the agent's possession of additional income.

Depending on the agent's social status and professional position, such activities and cover stories may vary significantly. For example, if an agent expresses the desire to purchase an automobile, in some cases it may be advisable to suggest that he acquire a used government vehicle at auction, priced significantly below market value. In other instances, it may be more appropriate to cover the acquisition of the vehicle as an installment purchase, with the agent paying in monthly increments. For those employed in engineering institutions, it is typically sufficient to attribute the additional official income to part-time translation work, which might amount to $15–25 per month, thus justifying intelligence compensation in the range of $100–150. She should also be advised not to disclose the true value of purchased items in conversation with acquaintances, instead quoting a reduced price.

An agent who knows a foreign language might take up translation work or tutoring in order to cover the income received from the intelligence service. It is often useful to recommend that the agent engage in two or three consulting contracts with private firms, allowing intelligence payments to be disguised as earnings. In many countries, governmental and charitable institutions frequently conduct lotteries and raffles. The agent may occasionally justify the possession of funds by claiming to have won a lottery ticket drawing.

The agent must be taught how to establish cover stories

and conceal his operational activity, particularly those related to intelligence collection or tasks linked to fulfilling the intelligence officer's assignments. For example, if the agent has acquaintances or friends working in locations of operational interest, a mutually developed legend should be created that explains the agent's interest in that location. The agent should also be advised to read books or articles covering the problem areas where the acquaintance works, thereby supporting his motivation for acquiring further information on the subject.

If the agent removes documents from work in order to hand them over to the intelligence officer, he must have a prepared legend ready for the event of an inspection of his briefcase or personal effects.

It is also necessary to cover through legend the agent's possession and storage of operational equipment, such as radio or photographic devices. If covering the storage of such equipment proves too difficult, then in some cases it may be preferable to conceal the very fact of the agent possessing these items. For storing this type of equipment, it is advisable to use special containers disguised as common household items.

Proper and skillful actions on the part of the intelligence officer in teaching the agent to cover and mask their operational activities, particularly those related to fulfilling intelligence assignments, will ensure sufficient tradecraft discipline in the agent's behavior and reduce the risk of detection by local counterintelligence authorities.

2. Training the Agent in the Execution of Intelligence Tasks

Training agents in the methods of intelligence work must be conducted systematically, using concrete examples during both the assignment and review phases.

When issuing tasks to any individual agent, their actual operational capabilities must be taken into account. Assigning an unrealistic or overly difficult mission may push the agent toward falsifying the information requested of him, or even lead to operational compromise. An agent, eager to complete a mission at any cost, may neglect risk assessment and act carelessly when attempting to acquire documents or classified information.

Incorrect assessment of the agent's capabilities is a serious shortcoming in working with them. It is not uncommon for an intelligence officer to disregard an agent's social or professional status. A person occupying a senior diplomatic or governmental position is often given tasks that are too minor or poorly conceived, due to the officer's excessive caution and underestimation of what the agent is capable of accomplishing. Conversely, the true operational potential of the agent is wasted.

For instance, a senior agent serving as an attaché in a in a capitalist country was primarily used for clarifying minor issues and obtaining background information upon request. He was never approached about exploiting the full operational potential linked to his official position. As a result, he never completed meaningful assignments and was not mentally prepared for cooperation with intelligence services.

During a meeting with the KGB Station Chief, a candid conversation occurred. The agent stated directly that he was dissatisfied with the way his work with intelligence was being conducted. He expressed unwillingness to transmit important political documents or their summaries unless his tasks were limited to minor inquiries, and only if meetings with the officer were arranged in a way that would not arouse suspicion.

At times, the intelligence officer fails to fully appreciate the capabilities of the agent. Some agents are underestimated, believed incapable of handling difficult assignments involving a certain level of risk. An agent may conceal the existence of valuable contacts from the officer and claim he is unable to access classified documents, even though his official position provides precisely such access. In order to fully exploit the agent's potential, the officer must clearly understand the agent's capabilities, derived from direct conversations and other available means.

When assigning a task to an agent, the officer must take into account not only the agent's intelligence potential but also their experience and the history of their cooperation with intelligence. A recently recruited agent should not be immediately given a complex task, even if he appears capable of completing it. Work with a new agent should begin with the assignment of simple tasks, increasing their complexity gradually as the agent becomes more established.

Even with an experienced agent, if a new assignment is particularly difficult, it is necessary to provide detailed instructions on how to proceed. The officer must discuss in advance the operational techniques best suited to that specific situation. A lack of thorough briefing can result not only in failure of the mission but also in the exposure of the agent.

In one Western European country, a stenographer-typist at the Dutch embassy, code-named "Lora", was successfully recruited. A local resident, she secured employment at the embassy through the assistance of a relative.

The KGB officer, Artem, to whom "Lora" had been assigned for handling, began clarifying the capabilities of the new agent. He determined that her duties included stenographing reports from the ambassador that were sent to the Ministry of Foreign Affairs in The Hague. Artem tasked Lora with supplying him the transcribed copies of those reports.

For two months, Lora met with Artem regularly, approximately once every two weeks, to deliver copies of the specified documents. Artem, upon receiving the materials, was not concerned with how she duplicated the documents, how they were removed from the embassy building, or where they were stored prior to the meeting.

After two months, Lora triggered an emergency signal to summon Artem and informed him that she had been dismissed from her position at the embassy under suspicion of espionage activity. During the debriefing, she recounted the following:

From the very beginning of her employment, she had noticed that she was being watched by the embassy's secretary. Later, it seemed to her that the surveillance had eased somewhat.

However, from time to time, the secretary would check on her work, entering the room where she typed and reviewed the stenographic records.

On one such occasion, the secretary entered the room just as she was completing her work and noticed that her typewriter contained not two copies, as was standard, but three. When asked why she was producing three copies instead of two, Lora replied that she must have misunderstood the instructions, having assumed the request was for three copies. The secretary, waiting until the last page was typed, then asked her to hand them over to finish typing and, in doing so, discovered that the remaining pages existed only in two copies. The secretary demanded an explanation from "Lora." Caught off guard and unprepared for such questioning, Lora became visibly flustered and could not provide a coherent response. The secretary then insisted on inspecting her belongings.

As a result of the search, the remaining pages of the third copy were found in Lora's handbag. It turned out that she had been slipping the pages into her bag one by one as they were typed, hoping in this way to conceal the existence of the third, unauthorized copy.

Lora attempted to explain her actions by claiming she wanted to become more familiar with the content of the ambassador's report, as she had a particular interest in the issue being addressed. However, she was unable to justify any specific personal interest in the matter.

Artem's error lay in the fact that, although he assigned Lora a task, he failed to properly train her in how to carry it out securely, with sound tradecraft. He assumed that Lora, young, inexperienced, and in need of meticulous instruction, would manage. He failed to inquire about the working conditions at her post. As it later turned out, there was no actual requirement to produce a third copy of the reports at all. The embassy did not require typists to submit extra carbon paper, and she could have passed the second copy to the KGB officer using carbon paper normally employed for the ambassador's reports.

Artem also failed to instruct Lora on how to conceal materials, how to store them before delivery, or how to act if they were discovered in her possession. As a result, a valuable agent was lost to the KGB.

When discussing methods and means of completing assignments with an agent, the case officer should take the agent's suggestions into account. Given the agent's familiarity with local conditions and the specific setting in which the task will be executed, the agent is often best positioned to identify effective and discreet solutions. The agent's initiative should be encouraged and developed.

However, the intelligence officer must not simply follow the agent's lead. Discussion of task execution methods between the officer and the agent is essentially an operational meeting, at which the final word must rest with the intelligence officer.

When assigning a task, the officer determines the deadline for its completion. In doing so, the officer must consider the urgency of the task, the actual time required by the agent to carry it out, the agent's personal qualifications and capabilities, the specific situation in which the task will be executed, and the logistics of organizing the next meeting.

An unrealistic deadline can push the agent to execute the task hastily or superficially. In some cases, it may lead the agent to disregard tradecraft protocols, which could in turn result in operational failure. Failure to meet deadlines undermines discipline and habituates the agent to irresponsibility. The intelligence officer is entitled to demand timely and precise execution of the task only if the time allocated was genuinely adequate.

The intelligence officer must think through the task carefully and articulate it clearly. The less precise a task is, the harder it becomes to monitor its execution. Agents, exploiting vague assignments or wishing to avoid complex tasks, may default to gathering only the most readily available information.

An intelligence officer must not limit their role to merely passing on assignments formulated at headquarters. Without detailed discussion of the substance of the task, misunderstandings can arise, as is evident from the letter of one agent to headquarters. The agent writes:

"At one of our meetings, I was assigned to visit several laboratories in the city of N. I asked which laboratory was of most interest to us. The operations officer I was meeting with replied: 'Check them all.' I traveled to city N, visited all laboratories except the ultrasound lab, whose director was away, and then returned home.

At the next meeting, it was revealed that the intelligence service was particularly interested in that very laboratory. Had I known this in advance, I would certainly have visited it.

Several weeks later, I was asked whether it would be possible to return to city N in order to inspect the ultrasound laboratory. Once again, I requested clarification on exactly what information was required, but received no reply. I visited several laboratories engaged in ultrasound research and compiled a report on my visits. The report, however, contained "many apparently useless items."

In this case, the lack of specificity in the assignment understandably frustrated the agent, who, not knowing exactly what was expected, had to spend significant time collecting information that could have been gathered far more efficiently with less effort.

Poorly defined, hastily issued assignments reduce agent motivation and, in some cases, cause the agent to withdraw from cooperation altogether. Carefully considered and specific assignments, on the contrary, engage the agent and increase the likelihood of successful completion.

For example, in one KGB Station, contact was reestablished with the agent "Dzherri," an ordinary official in one of the ministries. For several years, there had been no communication. In their first meetings, "Dzherri" expressed no objection to resuming cooperation, but he cautioned that he was unlikely to be of much use to the intelligence service, as he had no access to classified materials.

The case officer made no effort to conduct a thorough conversation with the agent to assess the nature of his duties at the ministry. Instead, he began issuing vague assignments related to the general collection of classified information. As a result of this poorly organized effort, contact was again lost. "Dzherri" tried to conclude his meetings with the officer as quickly as possible and eventually stopped attending meetings altogether. The Station Chief was only able to reestablish contact with great difficulty, but attempts to reactivate "Dzherri" as an intelligence source failed. Yet in the past, "Dzherri" had proven himself a positive contributor to intelligence operations. Thus, the cause of "Dzherri's" poor performance lay not in the agent himself, but in the case officer managing him. The Central Apparatus of the KGB made the decision to reassign "Dzherri" to another officer.

The first meeting between the new case officer and "Dzherri" ended without results, as the agent continued to insist that he had no ability to provide useful assistance. However, at the second meeting, the conversation turned to documents that "Dzherri" handled at work. It became clear that even official ministerial directories and reference materials kept on his desk could be of interest to intelligence. These were brought in by the agent, photographed, and returned to him. The Central Apparatus gave the documents a high evaluation, and this assessment was communicated to the agent.

Following this, the question arose: What other materials could "Dzherri" supply in the future?

The officer suggested that at their next meeting, "Dzherri" bring his current work folder, as well as several folders he had taken from his office shelves. He did so. The contents were photographed and assessed, allowing the Station to develop a precise picture of the types of documents passing through the agent's hands.

Based on this information, "Dzherri" was given specific assignments, which he successfully fulfilled. Gaining confidence and understanding what was expected of him, "Dzherri" began proactively selecting and supplying relevant materials.

The Station, in cooperation with "Dzherri," continued to discover new avenues of opportunity. As a result, "Dzherri" provided a substantial amount of valuable material. Thanks to proper management, both a valuable agent and a useful source of intelligence were preserved.

When assigning tasks, the officer must consider the agent's individual qualities. A timid and insecure agent must not be given tasks, the execution of which requires considerable courage. When assigning a task related to the acquisition of intelligence materials, it is necessary to explain to the agent which questions are primary and demand the greatest attention. It is important to warn him not to become distracted by secondary matters.

It is essential to require the agent to submit intelligence reports with the greatest possible completeness. He must present the essence of an event or a planned activity, indicate the time and place it occurred, provide information about individuals involved, specify the sources from which the data were obtained, and indicate the reliability of the information, as well as other necessary details for analyzing and evaluating the material received from the agent.

The case officer must not neglect the technical aspect of material delivery. The agent must be instructed on how to store materials in transit and how to destroy them if necessary (to discard, swallow, or burn). The case officer teaches the agent not to hand over intelligence materials personally to the officer, but rather to use secure communication channels. He demonstrates and explains how to use dead letter boxes and concealment containers.

When intelligence materials are being handed over in person, either to the case officer or a courier, the agent must be trained in how to do so while strictly adhering to tradecraft protocols. The agent must be informed that materials can be placed into an envelope, paper bag, newspaper, or item purchased from a store, and transferred in such a manner. This must be done without drawing suspicion, in the same natural way people pass along an ordinary letter, newspaper, or package.

At meetings, the officer may sometimes collect materials at the beginning, for instance, if it is necessary to review them on site, and at that point provide appropriate instructions to the agent. However, more often this is done at the conclusion of the meeting, once the case officer has conferred with the agent.

Proper task assignment and carefully structured briefings on how to fulfill the assignment will not guarantee success unless accompanied by regular and systematic oversight of how the agent carries out the case officer's instructions.

The main purpose of verifying an agent's execution of assigned tasks is to determine how completely the agent has fulfilled the assignment, who he spoke to, where and how he obtained the materials, and whether he conducted himself with adequate tradecraft and caution. Such verification enables the case officer to make timely corrections, expose deficiencies in the agent's performance, and mitigate their consequences.

The case officer must not allow the agent to fail at executing assignments without valid reasons. Each time an agent fails to fulfill a task, he must be required to provide an explanation. Naturally, the officer must act with tact, without offending or reprimanding the agent in a harsh tone, keeping in mind that the agent may have faced genuine circumstances interfering with the assignment. However, the agent must always feel personally responsible for the entrusted task. This type of systematic verification is desirable for all agents, regardless of their experience or functions in the intelligence field.

It is especially important to closely monitor the task performance of newly recruited agents. Some may not yet be able to carry out assignments with the necessary quality due to persistent feelings of fear. The officer must help the agent overcome this fear by instilling confidence, demonstrating that, by adhering to proper tradecraft and following the officer's recommendations, the risks of conducting intelligence work are minimized.

3. Strengthening and Expanding the Agent's Intelligence Capabilities

When working with an agent, the case officer must thoroughly understand all of the agent's intelligence capabilities and strive to use them as fully and effectively as possible. The officer must account for the fact that, due to insufficient training, some agents have a poor understanding of their own intelligence capabilities, while others, though aware of their potential, demonstrate excessive caution and do not always employ these capabilities fully in the interests of the intelligence service.

The case officer must not only be aware of the agent's capabilities, but must also take measures to prevent their atrophy. He must constantly work on expanding the agent's existing abilities and developing new ones; seek ways for the agent to penetrate new areas where valuable information can be obtained. In doing so, the officer must be guided by the overall needs of the intelligence service, not by a single operational line. Special attention must be given to identifying each agent's potential for work against the primary adversary.

When directing the agent toward the expansion of existing or development of new intelligence capabilities, officers typically recommend that the agent demonstrate appropriate initiative in his official duties. Performing such tasks, even if they involve some degree of operational risk for the agent, for instance, expressing interest in a certain topic unrelated to his official responsibilities, might raise suspicion among colleagues and eventually lead to the compromise of his existing intelligence access. Therefore, the risks of the proposed operation must first be assessed, and only if no doubts arise should the plan proceed.

To properly determine the intelligence capabilities of the agent, the case officer must carefully study the agent's status, know his standing in the workplace, how superiors view him, what relationships he has with colleagues, what access he has to documents of intelligence value, and whether he can carry out other intelligence tasks (such as making trips to dead letter box sites, serving as a courier, or functioning as the custodian of a communications point or safehouse address).

If an agent, for example, works as an official in a government ministry, the case officer must know exactly which documents the agent has direct access to through the nature of his work, and which documents he could familiarize himself with indirectly, without causing suspicion among colleagues or revealing the nature of his professional connections. In certain cases, it may be appropriate for the agent to initiate a series of officially justified actions that would allow him to familiarize himself with materials of intelligence interest. To this end, the agent may exhibit reasonable professional initiative, for instance, by proposing the study of a specific issue or the organization of an event. Such initiatives not only help broaden the agent's intelligence capabilities, but also grant access to valuable information under the pretense of officially sanctioned interest.

The case officer, during the professional cultivation of the agent, must also teach him how to properly conduct himself with supervisors, management, and colleagues to earn their trust and thereby solidify his own position.

Where appropriate, the case officer should advise the agent and jointly develop a plan for improving his professional qualifications (such as additional training or courses). To strengthen his professional standing, the agent must be diligent, disciplined, and modest. Neglect of the agent's official position by the case officer can lead to a loss of valuable access and, in some cases, the loss of the agent himself.

One example illustrates the consequences of such neglect: In a certain country, a valuable and experienced agent named "Heinrich" was part of the agent network. Due to specific circumstances, he remained unemployed for an extended period, despite being a highly qualified specialist in his field. The KGB Station, in coordination with the agent the Station developed several options to arrange employment for him at an institution of intelligence interest. This required substantial effort both on the part of the KGB Station and the agent himself, but ultimately the efforts proved successful: the agent secured a position at an organization of relevance to intelligence operations.

Being a capable individual, "Henrik" quickly adapted to his new position, and after some time began transmitting valuable documentary materials to his handler. The work carried out by "Henrik" was fully satisfactory to the case officer, and the case officer ceased paying close attention to the agent's position and professional relationships with colleagues and superiors.

Six months later, the agent reported to his handler that his immediate supervisor had drastically altered his attitude toward him: he was no longer entrusted with serious assignments and was being ignored. "Henrik" expressed concern that he might not be able to retain his post due to impending staff reductions.

Subsequent analysis of the agent's workplace relationships revealed that from the outset the agent had adopted an incorrect behavioral posture. Although he was by far the more qualified expert in his field than his immediate supervisor, "Henrik" consistently tried to underscore this superiority in order to establish authority among his coworkers. His supervisor, perceiving in "Henrik" a potential rival, abruptly changed his attitude, distanced him from sensitive tasks, and ultimately marginalized him. The Station, together with the agent, attempted to correct the situation and drafted an entirely different strategy for his conduct, but the moment had passed, and during the staff reductions, "Henrik" was dismissed from the institution.

Sometimes agents forfeit their intelligence potential because they inadequately conceal their progressive ideological views and consequently lose the trust of reactionary circles in which they are embedded for intelligence purposes.

If an agent possesses limited intelligence capabilities, then the efforts of both the case officer and the agent must be directed toward expanding existing ones or acquiring new ones. This can be achieved through various methods: establishing relationships in sectors of interest to intelligence; transferring the agent to a different job; selecting a suitable individual for the purpose of obtaining a particular post; or creating a cover employment situation that grants the agent access to institutions of intelligence interest (e.g., a newspaper publishing house, law office, or legal practice). The agent may also be trained in a specialization aligned with intelligence needs (for example, as a courier, acquiring skills as a driver, sailor, or pilot).

Successful case officer efforts to expand an agent's capabilities may be illustrated by the following example. Agent "Kane" was a mid-level bureaucrat at a government institution and had only limited access to valuable information. The head of the department was aware that "Kane" was a diligent, responsible, and competent employee and valued him. However, "Kane" lacked deep knowledge in the subject matter of his department, and his professional limitations were apparent to colleagues. As a result, his superior was unwilling to promote him to a position that would offer better access for intelligence work.

The case officer handling "Kane" regularly inquired into the agent's professional standing, his relationship with his superiors, and the internal dynamics of the department. He frequently sought information about the character of "Kane's" supervisor. After conducting a thorough analysis of the situation, the officer concluded that by upgrading "Kane's" professional qualifications and having the agent bribe his supervisor, it would be possible to secure his promotion to a position of significantly greater interest to the intelligence services, a position that happened to be vacant at the time. The case officer proposed that "Kane" pursue further training. "Kane" agreed, but on the condition that the KGB Station provide him with financial support, since the required studies would force him to forgo supplemental income.

Supported materially by the intelligence service, "Kane" applied himself diligently and, within a year, thanks to his natural aptitude, acquired the necessary knowledge. At a well-chosen moment, "Kane" spoke with his department chief about the possibility of being appointed to the position that was of interest to the intelligence service. In the course of the conversation, he discreetly implied to his superior that he would not remain in his current post indefinitely.

Some time later, "Kane" was summoned by the department chief, who showed him a prepared order appointing him to the desired position. It was clear that the department head expected a reward. Once "Kane" fulfilled this expectation appropriately, he was officially confirmed in his new position. In this way, the agent, thanks to correct management by the case officer, significantly expanded his intelligence capabilities.

The case officer must know the agent's contacts and patiently explain which ones should be maintained and which avoided. At times, it is necessary to dig into seemingly minor matters, such as mutual acquaintances of the agent and individuals of potential interest. In the environment of bourgeois society, evaluating such contacts can often be the first step toward cultivating new relationships, which must not be overlooked if the task is to expand the agent's intelligence capabilities by establishing new contacts.

An agent may develop useful connections through family members and leverage those connections in the interest of foreign intelligence. The following example illustrates how this can be accomplished. The wife of agent "Stenli" was sincerely devoted to his professional advancement and made efforts to help him secure a stronger official position. Acting on her own initiative, she built valuable relationships, which in turn helped reinforce her husband's status within his workplace and social circle.

The KGB Station was interested in ensuring that "Stenli" became acquainted with a prominent scientist, with whom, in the course of his normal duties, he had no direct contact nor any relationship with him. At a meeting with the agent, a plan was developed to establish this acquaintance through his wife. "Stenli," under the pretext of intending to advance his career, told his wife frankly that he needed this contact. She supported her husband's intention and handled the rest on her own. Through her social connections, she learned the address of the seamstress, owner of a fashion salon, who made dresses for the wife of the scientist. She began frequenting the salon and eventually became a regular client. It was there that she became acquainted with "Kobra," as the scientist's wife was codenamed.

Taking advantage of a conversation about fashionable dresses, the agent's wife recommended her seamstress to "Kobra" and began bringing her there. In this way, the agent's wife initiated contact and soon became a trusted acquaintance of "Kobra." From there, it was not difficult, under domestic and informal circumstances, for the women to introduce their husbands to one another.

The use of family members, in this case, the agent's wife and the wife of "Kobra", led to the successful recruitment of a new valuable agent.

The agent's personal connections often significantly exceed the opportunities available to him through his official position. The case officer must thoroughly study these connections to reveal the agent's potential intelligence value, to suggest how best to utilize them for completing intelligence assignments.

In one country, the intelligence service had an agent codenamed "Kristofor," who worked in the Ministry of Finance in a low-level position. Nevertheless, "Kristofor" provided economic intelligence of moderate value through access granted by the nature of his job. His case officer, who handled "Kristofor," regularly received from him reports on the country's economy. All case work with "Kristofor" focused on leveraging his official position to extract the most valuable intelligence possible.

At the beginning of the relationship with "Kristofor," the case officer inquired into his personal and professional connections. However, failing to find anything of immediate interest, he did not revisit the subject again.

Only two years later, in a casual conversation with "Kristofor," the case officer learned that his daughter, upon graduating from university, had taken a job at the American Embassy as a translator, and had already worked there for over a year. "Kristofor" had never informed the case officer about this, as he assumed that the intelligence service was only interested in economic information pertaining to his country, and thus never inquired over the course of two years about any changes in his personal life or among his relatives.

This example clearly demonstrates an oversight by the case officer, who failed to systematically study the agent's family circumstances, thereby depriving the intelligence service of the opportunity to timely utilize access to the American Embassy.

In carrying out work aimed at expanding the intelligence capabilities of an agent, the case officer must bear in mind that some agents are reluctant to pursue expansion of their operational potential: some are risk-averse, others try to avoid the additional difficulties and workload that such expansion entails.

Therefore, the case officer must guide this work in such a way that the agent becomes genuinely interested in the process, deriving either moral satisfaction or material benefit from it.

4. Specific Features of Working with Agents Recruited for Future Use

The purpose of working with agents recruited for use in the future consists primarily in preparing them to become dependable assets capable of conducting intelligence operations targeting highly significant objectives, particularly those into which penetration is currently impossible or extremely difficult.

Work with such agents requires that the case officer pay close attention not only to matters of ideological and political orientation but also to operational training, especially in preparing conditions for the agent's future deployment to areas of intelligence interest.

The key characteristic of an agent recruited for future use lies in the fact that, at the time of recruitment, such an agent usually has no immediate access to specific intelligence targets yet due to his personal qualities or social status, familial or other connections in political or business circles, the agent has a realistic prospect of securing employment in a post of interest to intelligence operations.

Training an agent recruited for future use, and overseeing his development, requires extensive, meticulous work, often complicated by the fact that the case officer, prior to the agent's employment in the desired organization, cannot yet obtain from him authentic intelligence materials. Nevertheless, such materials are exactly those the agent might later acquire under the case officer's direct guidance and instruction. Every case officer must think not only in terms of present operations but also focus on preparing agents for future missions.

It is known that the counterintelligence agencies of capitalist countries pay particular attention to "loyalty screening" of personnel employed in the most sensitive government institutions. These include intelligence and counterintelligence agencies, government chanceries, major ministries, research institutes, and laboratories engaged in nuclear energy, missile technology, and chemical or bacteriological weapons, precisely the kinds of institutions that present the greatest interest to the intelligence services of socialist states.

Personnel in such facilities are carefully selected and continuously monitored by counterintelligence. Conducting recruitment within such institutions is extremely difficult. Therefore, infiltrating these facilities is often possible only through agents recruited for future use.

If intelligence has recruited an agent who is currently a student at a privileged academic institution and who has influential relatives, then it is essential to begin working with him two or three years prior to graduation, focusing on helping him realize the potential for employment in a position of interest to the intelligence service. The agent should be advised to approach a certain relative and request their assistance in obtaining employment at the designated institution. If necessary, the agent should be encouraged to frequently visit that relative, to seek help from his father, mother, brother, or sister in resolving the employment issue.

All of the agent's actions must be guided by the intelligence officer, who must unwaveringly pursue the objective defined by the intelligence service at the time of the agent's recruitment. If the agent is studying a particular profession that would enable him to secure employment, after graduation, in a laboratory researching a significant scientific or technical problem, he should be encouraged in advance to show interest in the issue. He should be advised on choosing a thesis or dissertation topic, and be provided with the means to access consultations with highly qualified experts. To achieve this goal, the case officer may also use the agent's personal qualities and financial resources to "soften" individuals who could assist the agent in obtaining employment in a position of interest to the intelligence service, thereby granting access to valuable intelligence opportunities.

Working with a prospective agent network must be conducted through tasking assignments, the fulfillment of which helps to solidify the agent's relationship with the intelligence service. For instance, such agents may be tasked with collecting and documenting written background information (character references) about their friends, acquaintances, or close relatives. If the agent is a member of a political party or club, he may be asked to gather character data about the club's or party's leaders and key members. For completed assignments, if deemed appropriate, the agent may be issued monetary compensation and asked to sign a receipt, which also serves as a document affirming the agent's operational relationship with the intelligence service.

Experience has shown that from among such prospective agent networks, the intelligence service can prepare agents capable of securing positions in important government institutions .into the leadership of political parties, parliaments, and government circles, among prominent scientists.

5. Working with an Agent Recruited Under False Flag

Just as in the matter of political indoctrination of agent networks recruited under false flags, so too in the training of such agents in tradecraft and counter-surveillance, specific peculiarities arise, stemming from the fact that in such cases, the intelligence officer operates under a false flag while concealing their true allegiance. In this connection, special attention must be given to questions of tradecraft. It becomes necessary to disguise not only the conduct of intelligence operations from local counterintelligence, but also the fact that the agent is working on behalf of a socialist state.

Working with an agent recruited under false flag differs substantially from handling agents openly recruited on behalf of a socialist state. The challenge lies in the need to train the agent in tradecraft and cultivate him in accordance with the requirements of the intelligence service, while the officer himself does not present as an operative of a socialist country, but rather as a representative of a capitalist nation.

This requires a high degree of skill and inventiveness on the part of the intelligence officer from the socialist state, particularly in cases where the work is conducted under the guise of an ostensibly independent organization (a political party, commercial firm, newspaper, etc.). In such instances, direct instruction in intelligence methods may arouse suspicion, making the agent wary that he is dealing with an intelligence service rather than an innocuous entity.

The cultivation of agents recruited under false flag must therefore proceed along the lines of fostering businesslike traits and practical qualities essential not only by the agent's political orientation, but also by the flag under which he was recruited, the operational environment, the intelligence officer's position, and significantly by the nature of the personal relationship established between officer and agent. If, during the course of the collaboration, the intelligence officer succeeds in exerting political influence on the agent and effecting a shift in his outlook, it can be expected that the transition of the agent to direct cooperation with the intelligence service of a socialist state will not present significant difficulty. Consideration of the circumstances of the recruitment, the agent's personal qualities, the political climate, and other such factors is essential, they may either facilitate or, conversely, hinder the agent's direct transition to collaboration with our service.

In practical intelligence work, however, there are cases where the question of transitioning the agent to direct cooperation with the intelligence service of a socialist state is never raised. This typically occurs in situations where operational conditions or the agent's position preclude such a transition or where it may endanger the agent's viability altogether. In some cases, the very nature of the agent's assignments and his specific role make it possible to use him only under a false flag.

6. Methods of Influencing the Agent

The intelligence service employs various methods of influence on the agent network, both material and moral in nature. These include both incentives and disciplinary measures. The choice of method is determined primarily by the underlying reason that necessitates exerting influence over the agent.

In this process, the intelligence service must consider the agent's social and professional standing, his political and cultural development, his personal qualities, and the internal motivations that compelled him to work for the service, in particular, his interest in maintaining his covert relationship with the intelligence apparatus. If the agent cooperates with .the intelligence service on a material basis, then the greatest effect will be achieved through incentives of a material nature. However, when dealing with an agent who is not materially motivated in their collaboration with the intelligence service, such incentives will not yield the desired results. To select the proper type and degree of influence, it is essential to take into account the agent's psychology, temperament, character traits, and material living conditions.

An agent completes tasks more willingly when he senses constant attentive regard for his work, a fair assessment of his efforts, and acknowledgment, either verbal or tangible. When encouragement is applied correctly and in proportion, it generates in the agent a sense of satisfaction from working with the intelligence service and promotes initiative in fulfilling assignments. However, excessive praise without corresponding results can foster arrogance in the agent, who may then direct grievances at the intelligence officer. Correcting negative traits that develop in the agent's character or behavior as a result of unearned praise can prove extremely difficult. Therefore, from the outset, reward must be proportionate to the agent's actual contributions.

In the operational practice of the intelligence services of socialist states, a variety of incentive measures are used, depending on the importance of the services rendered by the agent. These may include expressions of gratitude, gifts, financial rewards in the name of the case officer or the Center, meetings with senior personnel from the intelligence organization as a sign of recognition, assistance with career advancement in institutions or enterprises, and the facilitation of the agent's naturalization as a citizen of a socialist state. The latter type of reward is particularly reserved for agents who, by origin or conviction, are linked to socialist nations and have expressed a desire to "return to the homeland."

Some types of encouragement require authorization from governmental bodies; others require approval from the central apparatus of the intelligence service, the third, approval by the KGB Station Chief with subsequent notification of the central apparatus. For example, the matter of issuing an expression of gratitude may be resolved by the KGB Station on site, if the central apparatus provides a positive assessment of the agent's performance. However, approval from the central apparatus is required as the basis for such an action. Similarly, incentives such as granting citizenship or awarding state honors require decisions by governmental authorities and may only be applied to agents with particularly distinguished achievements. Such measures are employed very rarely.

Since the agent continues to reside in the country, documents relating to citizenship or awards are typically only shown to the agent. These are retained either at the KGB Station or within the central apparatus of the intelligence service. If the situation permits, it is preferable for the presentation of these documents to occur not in the course of an ordinary meeting but in a ceremonial setting.

On occasion, an agent may accept citizenship of a socialist state but never utilize the rights it provides until the end of their life. However, the knowledge that he possesses the high status of being a citizen of a socialist country or has received state honors from that country serves to encourage him in his dangerous work.

A meeting between the agent and a senior intelligence officer to convey gratitude, to present awards (if the operational environment permits), should be arranged accordingly. In addition to its emotional and motivational significance, such a meeting also has a beneficial effect, since the intelligence officer may use the occasion to deliver specialized instruction to the agent. Such meetings are sometimes organized when the agent has an opportunity to travel to a socialist country as part of a delegation.

The most commonly applied form of influence on an agent remains the evaluation of his work, delivered either directly by his handler or on behalf of the central leadership of the intelligence service. For an agent cooperating with the intelligence service on an ideological-political basis, the significance of such an assessment is especially great. In evaluating the agent's work, the case officer must be objective and take into account a number of factors which may determine whether the agent's performance was good or poor.

Occasionally, an agent, due to circumstances beyond his control, may be unable to complete a task, despite significant effort. If the case officer, failing to understand the reasons behind the nonfulfillment, expresses dissatisfaction with the agent's work, this may not only fail to motivate the agent to increase his activity but could produce the opposite effect. If the case officer is unfair or biased in his evaluation, the agent may become disillusioned, exhibit passivity, and fail to apply the necessary effort to accomplish more complex assignments, particularly those involving heightened risk.

In the practice of intelligence work, there are instances when information acquired by the agent is not of immediate value, because it has already been obtained by another source. Naturally, such circumstances should not influence the assessment of the agent's effort. On the contrary, the availability of corroborating data from other sources strengthens the credibility of the agent's report, increases its value, and once again attests to the agent's conscientiousness. At the same time, agents should not be praised excessively or prematurely, as this may lead to arrogance and a decline in performance quality.

In special cases, as exceptions, when an agent deliberately refuses to carry out the tasks assigned by the intelligence service, it may be deemed appropriate to apply disciplinary measures against such an untrustworthy agent, including threats of exposure, and blackmail. This may involve the use of compromising materials, not related to the agent's intelligence activity, but rather threats to expose evidence of past personal or professional misconduct.

Resorting to threats is permissible only in extreme cases, when the intelligence service possesses serious compromising materials or other leverage allowing it to hold the agent under control; but of course, threats should never be brought to the point of actual execution.

The most common form of material incentive is the payment of the agent's work. This is practiced not only in relation to agents working for material interest, but also to those collaborating with the intelligence service on an ideological-political basis. Besides being a form of encouragement, payment for the agent's work also serves to secure the agent's loyalty to the intelligence service and to increase his productivity. Moreover, proper application of financial compensation serves as an additional tool for fostering the agent's professional qualities.

When using monetary remuneration as a means of stimulating and consolidating agent networks, it is important to understand that issuing an unjustifiably large sum corrupts the agent, fosters consumerist and mercenary attitudes, and reduces his effectiveness. On the other hand, underpayment generates dissatisfaction and resentment, which not only may affect the agent's performance but also weaken his will to carry out assignments.

The determination of payment amounts in each specific case must be based on the agent's work quality, the complexity of the tasks performed, the living conditions of the local population, and the service-related, social, and financial situation of the agent. When using money as an incentive, the case officer must remember that prudent handling of public funds is one of the essential elements of a case officer's sense of duty to the Motherland.

In the operational practice of intelligence services of socialist states, several types of monetary compensation for agents were established, applied according to the specific conditions of the agent's work and personal life.

Monthly compensation consists of a fixed sum paid to the agent each month. The amount, as previously indicated, depends on the quality of the agent's work, the subsistence minimum in the country, and the agent's social standing. This form of compensation is typically used for auxiliary agent networks (such as keepers of safehouses, addresses, couriers, and cutouts), where use of alternative compensation systems is impractical for understandable reasons.

One-time remuneration refers to a payment for the successful completion of a specific task by the agent. The criteria for determining the size of this compensation include the value and usefulness of the information or services provided to the intelligence service, the complexity of the assignment, and the risk undertaken by the agent in its execution. These same factors, namely, the agent's material situation and the specific living conditions in the country, are also taken into account. This form of compensation promotes the agent's motivation and performance.

Mixed (combined) compensation systems combine monthly payment with one-time rewards. Under this system, the agent receives a fixed monthly sum, and in addition, is granted extra monetary compensation for particularly valuable material or information, or for the successful completion of particularly important tasks. The amount of this additional payment is determined by the value of the material and the difficulty of the task performed.

The method of compensation is selected by the KGB Station Chief depending on the agent's personal characteristics, his value, and the expected impact of a given form of payment on the agent's productivity. When determining the type and amount of compensation, it is also necessary to consider that the agent may, due to counterintelligence concerns, have to abandon one post in favor of another, thus affecting the feasibility of his operational position. For instance, in one country, intelligence used a typist from a government office as a covert mail courier; however, the individual held a visible public position and was in danger of being promoted to a higher-paying position. To keep him in his current post without compromising his financial interest, the KGB Station added the difference in salary between the courier's position and the promoted one to his existing remuneration.

The various forms of compensation described above each have their specific features, which affect the agent network in different ways depending on personal qualities and the degree of loyalty to the intelligence service. For instance, monthly compensation granted to an unreliable agent cooperating solely for material reasons may result in him delivering only insignificant information, the acquisition of which poses no effort or difficulty. On the other hand, a system of task-based one-time remuneration for such an agent may serve as a stronger incentive for more productive work on behalf of the intelligence service.

Agents cooperating on an ideological-political basis often reject financial compensation altogether, even when in material need, believing that taking money would introduce an element of mercenariness into their work. In such cases, the intelligence officer must explain that the funds offered are not a form of payment, but are instead intended to improve the agent's living conditions, cover his operational expenses, and, consequently, foster improved conditions for intelligence work. For these agents, it is appropriate to issue a monthly stipend under a suitable pretext, such as improving living standards or supporting health-related needs.

Attentiveness to an agent's personal circumstances and expressions of interest in his well-being stimulate a stronger desire to cooperate. It is beneficial to encourage agent performance through gifts, such as birthday presents for their children, small items for holidays or the New Year, and other festive occasions. In doing so, it is essential to correlate the value of such gifts with the agent's social standing and material conditions.

When selecting gifts, one should avoid purchasing unique items that attract attention. It is also advisable to avoid buying goods from stores where the buyer's affiliation with a particular institution in a socialist country is known. Gifts must be selected and delivered in such a way that it becomes impossible to determine their source. In certain cases, to preserve tradecraft, it is better to provide the agent with a sum of money and specify the item the intelligence officer recommends for the agent or his wife, allowing the agent to purchase it himself.

As a rule, agents should be compensated in the currency of the country in which operations are being conducted, using bills that can be exchanged at banks without requiring registration of the name or address of the bearer, and without the need to present identification. Agents must not be paid in freshly issued banknotes or bills from newly arrived couriers, especially if these come from socialist-state institutions or are brought across the border. Banks in capitalist countries often issue newly minted currency with recorded serial numbers, which are tracked by counterintelligence. Therefore, the presence of such banknotes in an agent's possession may lead to exposure.

A receipt for compensation rendered for services provided to the intelligence service, though a compromising document, plays an important role in anchoring the agent's connection to the service. Frequently, the agent signs such a receipt with great reluctance, trying to avoid it under various pretexts. This reluctance is typically explained by the agent's fear of placing a compromising document in the hands of an intelligence officer. Therefore, it may sometimes be necessary to forgo the receipt, at least the first time, and revisit the issue later. In certain situations, it may be necessary to refrain entirely from requesting a receipt, particularly if the agent is motivated by operational rather than ideological considerations. Nonetheless, from a procedural standpoint, obtaining a receipt is preferred, as it provides the necessary documentation for accounting purposes related to agent compensation.

There is no need to insist that the agent always write explicitly in the receipt that the remuneration was for completing intelligence assignments. It is sufficient for the agent to indicate the amount received and sign using a fictitious name. What matters is not the specific wording of the receipt, but the psychological impact on the agent of knowing that such a receipt exists in the possession of the intelligence service.

PART IV

VETTING AND ONGOING ASSESSMENT OF AGENTS

Introduction

The intelligence service of a socialist state continuously monitors and evaluates its agents, regardless of the circumstances of their recruitment or the length of time they have been carrying out intelligence tasks. This necessity arises from the reality that an agent's status in society, at work, and within the family changes over time, and these changes are accompanied by shifts in mood and views that can broaden or diminish their operational usefulness.

Under the influence of these factors, an agent may either grow closer to the intelligence service of the socialist state or, conversely, drift away from it.

Ongoing study and vetting of agents allows the intelligence service, first and foremost, to better organize the agent's tasks and most effectively exploit their operational capabilities. Secondly, it enables the identification of potential moles, disinformation agents, and informers who may have infiltrated the agent network.

Thorough study of the agent and his surroundings allows for early detection and prevention of conditions that might lead to the agent's compromise. Among the agent's friends and acquaintances, individuals may appear who are suspected of having ties to foreign intelligence services, criminal elements, or otherwise raise concerns. As a result, the agent may fall under counterintelligence surveillance and become the subject of an active file, which in turn could lead to exposure of his cooperation with the intelligence service of a socialist state.

Equally dangerous to the agent is continued contact with representatives of local progressive organizations. Agents often maintain such ties, particularly if they were formerly active with these organizations, despite explicit prohibitions from intelligence personnel. Some continue these contacts out of habit or a desire to maintain old relationships. It is the task of the intelligence officer to identify such risks in a timely manner and warn the agent about the dangers of maintaining those connections.

Agents must also be warned against visiting places known to attract criminal elements (such as nightclubs, and certain cafés), as well as establishments frequently attended by members of progressive organizations, since these locations are commonly under police and counterintelligence surveillance.

Comprehensive study of the agent enables the detection and optimal use of his intelligence potential, as well as identifying paths for expanding that potential.

Counterintelligence and intelligence services of capitalist states are constantly attempting to infiltrate the agent networks of socialist intelligence services through carefully placed moles and disinformation agents. Counterintelligence uses these moles to uncover elements of the intelligence network, to expose personnel affiliated with intelligence services, to identify methods of tradecraft, and to spread disinformation or derail operational plans.

To penetrate the agent networks of socialist states, capitalist intelligence and counterintelligence agencies deploy their own agents in the form of "planted assets," with the intent of recruiting from within KGB Stations or passing along intelligence on identified agents. In some cases, they deliberately attempt to pass off compromised or double agents to the intelligence services of socialist countries.

For these purposes, adversary counterintelligence services deploy a variety of operational ruses, strategies designed to attract the attention of our intelligence to agents deliberately offered by the adversary. In such ruses, the planted agent often presents himself as a politically sympathetic countries and ostensibly willing to provide assistance to our intelligence services. Quite often, such planted agents are introduced to us through front organizations or based on recommendations from individual progressive figures or organizations, or even from citizens of socialist countries.

It must be noted that in recent times, direct and crude attempts to plant agents, such as moles presenting themselves at socialist institutions with unsolicited offers of services, have become relatively rare. These services, moreover, often lack any genuine intelligence value. A more sophisticated strategy by capitalist counterintelligence services involves complex long-term operations, sometimes unfolding over the course of years. For this purpose, special organizations of an ostensibly progressive nature are created, which go on to publish journals, newspapers, and so forth.

Typically, the adversary's counterintelligence attempts to reintroduce individuals previously exposed as agents of the socialist intelligence services. These individuals may have been compromised for a variety of reasons. Sometimes, they turn traitor under pressure, fear, or a change in their political convictions.

Occasionally, agents within the agent network of a socialist intelligence service are found to be simultaneously collaborating with capitalist intelligence services, i.e., functioning as double agents. These individuals usually conceal their collaboration with other services and, motivated by financial gain, deliver duplicate materials to multiple services.

The presence of moles, traitors, and double agents within the agent network can cause severe damage to the intelligence service. Therefore, it is vital to conduct ongoing assessment and verification of agent personnel in order to identify and remove such elements.

In the practical experience of socialist intelligence work, agents are sometimes encountered who attempt to pass off fabricated or plagiarized materials as genuine intelligence, presenting them as secret or exclusive, while citing dubious or unverifiable sources. In some cases, agents are inserted into an already-established agent network, consuming funds intended for transmission to another agent or earmarked for carrying out operational tasks. There are also agents who demand excessive compensation for information allegedly acquired through considerable effort or expense.

Alongside such cases, there are agents who deliberately evade the performance of intelligence assignments. Timely identification of such agents allows the intelligence service either to sever ties with them or to find means of compelling them to fulfill their duties conscientiously.

The systematic and thorough study and verification of agents must be conducted throughout the entire duration of their operational relationship. This principle is one of the fundamental conditions for the effective functioning of the agent network of the socialist intelligence services.

In addition to several broadly employed methods for studying and verifying agents, developed from operational practice, external records of an agent must also be checked not less than once a year against the operational registry maintained by the Center. Such verification is essential because the records in the operational registry are continuously updated and new information about an agent may arrive from other sources, such as other KGB Stations or from military intelligence.

The primary methods employed by the intelligence services of socialist states for studying and verifying agents include the following:

1. Studying the Agent Through Personal Contact

Personal interaction between the case officer and the agent is of paramount importance in the study and verification process. Such contact allows the officer to form a direct impression of the agent, to correctly interpret and explain their actions, to correct errors, and to exert the necessary influence.

Through personal meetings, the case officer is able to assess the agent's political views and ideological convictions, as well as their personal behavioral traits; to clarify specific aspects of the agent's biography; and to inquire into his professional and family affairs, his material circumstances, and to evaluate his intelligence potential. When studying the agent through personal contact, the case officer must pay close attention to various, even seemingly insignificant, details: to the agent's views, expressions, and moods; to his remarks and the manner in which he demonstrates particular interest in certain topics; and to any inconsistencies or contradictions in his statements regarding himself or his contacts. Such data must be systematized and incorporated into the agent's profile, compared with information he has previously provided, whether verbally or in written reports.

As a result, the case officer can obtain not only a fuller understanding of the agent but also potentially uncover deception or deliberate attempts to mislead the intelligence service. In this way, it is sometimes possible to identify moles, double agents, or individuals under the influence of hostile counterintelligence services who have infiltrated the agent network.

Personal interaction with the agent also enables the case officer to identify signs characteristic of moles who have been planted in the agent network by counterintelligence. Such moles often treat tradecraft matters casually, are unafraid of being compromised, display excessive interest in the case officer's identity and personal circumstances, and show little concern for intelligence work itself. They may display indifference or disdain toward completing tasks assigned by the service.

A mole inserted by enemy counterintelligence will often accept difficult assignments without hesitation, yet fail to complete them, or do so in a deliberately negligent manner. Such conduct often stems from their confidence that they will not face consequences for collaborating with the intelligence service of a socialist state, since they operate with the knowledge and protection of the hostile counterintelligence service.

These types of behavioral indicators require the intelligence officer to respond with rigorous verification of the agent by all available means, even when no suspicious behavior is immediately apparent. .the agent cannot be studied through personal contact alone. One must also consider that counterintelligence often attempts to insert moles into the agent network who have been thoroughly trained and briefed on how to behave so as to avoid raising suspicion.

Naturally, personal interaction alone is insufficient for checking and evaluating an agent. Personal impressions can be biased or inaccurate. Typically, identifying moles, double agents, or dishonest agents is achieved through a combination of verification methods. The unreliability of personal contact as a sole method stems from the fact that, like any person, the agent may wish to present himself in a more favorable light and conceal his mistakes or shortcomings. The agent may hide from the case officer any shifts in political orientation, any change in attitude toward the service, or his attempts to obscure ties with the intelligence service from his family and acquaintances. Thus, one cannot assess an agent solely on personal impressions; it is necessary to acquire and carefully analyze information about him from additional sources.

2. Cross-Agent Verification

This method of verifying and assessing an agent involves assigning a task to another agent, one who has some relationship or contact with the environment surrounding the agent under review. That agent is tasked with gathering information about the subject's character, activities, contacts, and other facets of his personal or professional life, resulting in a double-blind assessment. When deciding whether to use this method, it is essential to first determine how appropriate it is in the given situation.

Cross-agent verification must always take into account the potential for reciprocal exposure. Since any assessment of one agent by another inherently carries a degree of risk, the KGB Station Chief must evaluate not only the feasibility but also the appropriateness of employing this technique. In some cases, the Station Chief may identify an agent who is in a position to discreetly collect useful information about the subject. Typically, such opportunities are available not only to agents from the immediate circle of the subject under verification (relatives, close friends, acquaintances, colleagues), but also to those who are in the same political party, sports or religious organization, work in the same institution, or reside in the same building.

In cases where the agent's operational role is of particular importance to the intelligence service (for example, involvement in measures to influence certain spheres of public life in the target country), the KGB Station may deliberately recruit an agent from the subject's environment for the specific purpose of external verification, or it may make use of an already recruited agent.

Cross-agent verification as a double-blind method is employed primarily when there are reasons to question the integrity and loyalty of the agent under scrutiny, or when the agent is suspected of being compromised. In order for the verifying agent to act without arousing suspicion in the subject or revealing to him that he is being checked, the case officer must give detailed instructions. These must specify what individuals or circumstances are of interest, and what kind of behavior is to be observed in the agent's workplace, home life, political views, and known contacts. The verifying agent may also be tasked with discreetly making inquiries about the agent's background, covering up the fact that another agent is being used to monitor him.

The use of other agents for evaluating the veracity of information provided by (or directly surveilling) other agents requires the case officer to exercise utmost caution and ingenuity.

3. Analysis and Comparison of Agent-Provided Materials

Regardless of what materials or intelligence the agent delivers, or how long the agent has been in cooperation with the service, or how trusted he is by the intelligence apparatus, all materials received from the agent must be subject to critical assessment and analysis. Through analysis of such materials, the intelligence service is able to guide the agent's operations with precision, clarify certain key questions, and timely point out any deficiencies in the agent's performance. Moreover, a thorough analysis and comparison of the information received from the agent often allows for the timely detection of signs indicating unreliability, identification of moles, exposure of agents working for foreign intelligence services or counterintelligence agencies, or the detection of individuals who may be transmitting disinformation to the intelligence service of the socialist state, or even fabricating so-called "intelligence data."

The receipt of both documented and undocumented

reports from the agent, whether routed through the KGB Station or through the central intelligence headquarters, when correlated with information received via other agents or through the use of legal (i.e., overt) sources, provides an important basis for validation. Comparative analysis of the agent's reporting may lead to the conclusion that the agent is not independently acquiring the intelligence he claims. Through careful verification of the context in which the documents or reports were obtained, and by comparing their contents with verifiable facts and official records, it is often possible to establish whether the agent is fabricating "intelligence data."

For example, in one Central European country, a mole was successfully exposed within the agent network of a socialist state's intelligence service. This agent, positioned as a stenographer in the meetings of a ministers' council, submitted a report claiming that at one such meeting, Minister X, identified as a representative of the Christian Democratic Party, had participated. Some time later, a short public communiqué appeared in the press summarizing the meeting of Christian Democratic Party delegates from various countries, including Switzerland, and in this communiqué Minister X was listed as a participant. When comparing the date of the meeting in Switzerland with the date of the Council of Ministers' session, which the agent had reported, it was established that both events had occurred simultaneously. Thus, Minister X could not have been present at the Council meeting. This discrepancy served as the first signal raising doubts regarding the authenticity of the documents provided by the agent.

Subsequently, through additional intelligence capabilities of the KGB Station, it was established that the external formatting of the documents presented by the agent as official deviated significantly from the standard formats used in the apparatus of the Council of Ministers. This finding further reinforced suspicions held by the Station.

The content of the documents submitted by the agent was then subjected to detailed analysis within the central intelligence apparatus. Upon comparing the agent's earlier materials with the reports from another agent on the same issues, it was revealed that the documents contained several contradictory statements.

Additionally, the specific nature of questions allegedly discussed at the Council of Ministers' meeting, and the phrasing used, led to serious doubts about their authenticity.

Based on the thorough and comprehensive analysis of all materials received from the agent, the intelligence service concluded that it was dealing with an unscrupulous agent who had embarked on a path of fabricating intelligence materials in order to extract large monetary rewards from the service. As a result, it was decided to conduct a formal debriefing session with the agent aimed at exposing him as a deceiver. The course of the discussion fully confirmed the suspicions of the intelligence service. Pressed to explain himself, the agent was forced to admit that he had lost his ability to obtain official documents ten months earlier but, not wishing to forfeit financial compensation, had resorted to fabricating materials.

Nevertheless, the receipt of disinformation via an agent does not always indicate betrayal. In certain cases, enemy counterintelligence agencies maintain contact with a particular agent of the socialist state intelligence service not in order to re-recruit him, but to exploit that connection for disseminating disinformation materials through him.

It must also be taken into account that sometimes even genuine official documents may, due to their content, contain elements of disinformation and thus lead the intelligence service to question their authenticity. This often occurs in documents issued by embassies of capitalist states and sent to their Ministries of Foreign Affairs. The content of such documents may be based on dubious or unreliable sources. Authors of these documents may, intentionally or not, present a biased interpretation of facts. This possibility must be considered when evaluating the agent's submissions, to avoid mistakenly suspecting him of provocative intent, particularly in cases where the content reflects the official position of other recognized authorities.

The study of an agent through analysis and cross-verification of the materials he delivers is conducted by both the central apparatus of the intelligence service and the KGB Station.

4. Assigning Test Tasks

The practice of intelligence work has shown that posing a test or specially designed verification assignment can be an effective method for checking and studying an agent. For example, an agent might be tasked with acquiring information on an issue already known to the intelligence service. However, such a task should only be assigned if the agent demonstrably has a real and feasible opportunity to obtain truthful and verifiable information on that matter. Otherwise, if the agent provides information that contradicts what the intelligence service already knows from other sources, this cannot yet be taken as sufficient grounds to suspect the agent of dishonesty.

For the purposes of agent verification, the KGB Station may assign him a task which is clearly impossible to accomplish within the designated timeframe, or a task which obviously exceeds the agent's operational capabilities. An agent who readily accepts such a task and later falsely reports its completion on time, without any issues, should raise suspicion and concern in the mind of the case officer.

Another method for verifying an agent involves handing him a package purportedly for transfer to another individual or for temporary storage. The package contains materials in a special wrapper designed so that any tampering would be easily detected by the intelligence service. Below is an example illustrating such a verification method:

The intelligence service once decided to test an agent ("Lodo") whose behavior had raised doubts. A covert drop (dead letter box) was prepared with a package labeled as containing "intelligence materials." The agent was instructed to retrieve these items from the drop and deliver them to the case officer. Upon execution of the task, it was discovered that certain security seals on the package had been broken, clear evidence that someone other than the agent had accessed the contents. Additionally, surveillance of the drop site revealed that the area was under covert observation by hostile intelligence, indicating a breach.

It should be noted that in order to identify planted agents through such test assignments, careful preparation of active measures is essential. These assignments must be designed to provoke the adversary into action. Their content should interest counterintelligence services and compel them to act, thereby exposing themselves in an attempt to acquire valuable intelligence materials. Such materials, they believe, could be used to detect the operations of socialist intelligence networks or fuel provocations and propaganda, including the orchestration of show trials.

5. Verification through Surveillance and Observation Posts

In the context of foreign operational work, the organization of surveillance over an agent, the setup of observation posts, searches, and the interception of correspondence is associated with significant technical difficulty and typically requires that the KGB Station maintain a proven and reliable agent network. Nevertheless, these measures, when carried out for verification purposes, yield highly effective results, and in many cases, their use is essential.

Through external surveillance, it is possible to detect:

- the agent's contact with suspicious individuals;
- the agent's visits to apartments known to belong to local police or counterintelligence;
- the agent's entry into buildings that house our local intelligence presence or enemy counterintelligence services.

Surveillance may reveal that the agent under scrutiny was followed to a meeting location with an intelligence officer of the adversary. Or, conversely, that the agent did not travel anywhere, contrary to what he claimed when reporting to his handler, claiming, for example, to have visited another city to meet a "friend." The intelligence service, by organizing surveillance, can uncover a wide range of details about the agent's life and activities.

Here is an example of agent verification via observation:

Within the agent network of one KGB Station, there was an agent codenamed "Douglas," about whom suspicions had arisen. The Station developed a plan for verifying "Douglas," which was implemented as follows:

A meeting was scheduled with "Douglas" for 8:00 PM. During the meeting, the intelligence officer informed "Douglas" that the next day, at 7:00 AM, a trusted individual would be traveling to another city. This individual was to be tailed and was allegedly in possession of a package containing classified materials destined for intelligence transfer. Since on this trip this train would only stop in the designated city for 10 minutes, meaning the meeting with the courier could only take place at the station platform. The intelligence officer explained to "Douglas" that, as an officer of a socialist state's official institution, it would be difficult to justify his own presence at the station at such an early hour. Therefore, he asked the agent, by way of exception, to meet the courier, accept the package from him, and deliver it to the officer later that same day. "Douglas" readily agreed to carry out the assignment.

The intelligence officer then described in detail the appearance and identifying features of the man whom "Douglas" was to meet, indicated the train car number, and informed "Douglas" that the man would appear on the platform and, after exchanging the agreed-upon recognition phrase and receiving the counter-sign from "Douglas", would hand him the package.

After the meeting concluded, "Douglas" was placed under surveillance by the Station. It was established that, after parting ways with the officer, he took several detours before entering a telephone booth, where he made a brief call. He then did not return home, but instead headed to a park on the opposite side of town, where he soon met with a middle-aged man. After a short conversation, they parted ways, and the man quickly entered a waiting vehicle parked near the park, while "Douglas" returned home.

The next morning, the surveillance team was stationed at the railway station and, shortly before the train's arrival, observed two suspicious individuals with suitcases who were clearly attempting to blend in with the arriving passengers. "Douglas" appeared on the platform moments before the train's arrival. As the train pulled in, "Douglas" approached the designated train car, as had been explained to him by the officer, and carefully examined the passengers disembarking. Meanwhile, the two suspicious men, who had been observing the platform, watched "Douglas" the entire time without boarding the train themselves. When they saw that "Douglas" had neither approached nor made contact with anyone, they left the platform after the train departed. "Douglas" quickly made his way to a payphone, placed a call, and only afterward returned home.

In another instance, through external surveillance, the Station uncovered the agent's policy of deceiving the intelligence service. The agent "Nathan" claimed that he received scientific-technical information from a friend who lived in another town several hours away. "Nathan" insisted that the intelligence service allocate additional funds as compensation for this individual's assistance and also demanded reimbursement for travel expenses related to his supposed visits to the town, which he claimed to have made twice a month on Sundays.

The Station established that on Sundays "Nathan" was under surveillance for an entire month and did not leave the city even once. Nevertheless, during meetings with the case officer, he continued to provide materials while maintaining that they were brought to him by his friend from the other town.

Under pressure from the intelligence officer, "Nathan" eventually admitted that his story about the friend in another town was fabricated and that the so-called scientific-technical information was, in fact, largely compiled from publicly available scientific-technical bulletins. These were intended for institutional use and had been obtained from the research institute where his wife's brother was employed.

In agent verification work, significant assistance can be rendered by conducting inquiries at the agent's place of residence or employment. Such checks allow for determining not only the agent's family and financial circumstances, but also his lifestyle, typical social circle, and, to some extent, details of his past. Sometimes, in the course of agent verification, it is advisable for the intelligence service to carry out operations involving the agent's correspondence.

In specific cases, an agent may also be checked through the use of technical surveillance methods, such as wiretapping his conversations with contacts, monitoring his telephone communications, and verifying his correspondence or activities related to dead letter boxes.

These, then, are the principal methods for the study and verification of agent networks. However, they by no means exhaust all the possibilities available to intelligence in this domain. Depending on local conditions and the specific characteristics of the agent, a wide variety of methods and techniques for studying and verifying the agent network may be developed.

The task placed before the intelligence officer is to demonstrate the utmost ingenuity and initiative in this matter. In most cases, when conducting the study and verification of the agent network, an effort is made to combine several of the above-mentioned methods.

PART V

INTERACTION BETWEEN INTELLIGENCE OFFICERS AND AGENTS

Introduction

In working with each agent, the intelligence officer solves a critical and highly responsible task: the preparation of a reliable and loyal agent of the socialist state's intelligence service, one capable of competently carrying out the mission assigned to him. This task can only be fulfilled if the proper relationship between the intelligence officer and the agent is established.

When working with an agent, the intelligence officer must always take into account the agent's individual traits. On the basis of practical experience across the intelligence services of the socialist commonwealth, several key principles have been developed that must guide the officer in managing contact with an agent:

Relations with the agent must be built on a strictly professional and conspiratorial basis, guided by the needs of the intelligence mission.

In the relationship with the agent, the leadership role must remain firmly with the intelligence officer.

In working with the agent, the intelligence officer must be attentive, tactful, and objective.

Constant vigilance and caution must be exercised in dealing with the agent.

The intelligence officer must maintain continuous oversight of the agent's work and behavior.

Let us now examine the substance of each of these principles in more detail.

1. Establishing the Relationship with the Agent

The relationship between the intelligence officer and the agent must be built on a strictly professional basis, that is, it must be guided solely by the interests of the intelligence service. The intelligence officer is required to shape his relationship with the agent in such a way that the agent perceives him not as a personal acquaintance, but as a representative of the intelligence service with whom he is cooperating. Otherwise, the officer will find it difficult to ensure the agent's strict adherence to tradecraft, contact discipline, and operational performance.

The intelligence officer must not allow personal contact with the agent that is unrelated to the operational mission. He should not become acquainted with the agent's family or, for example, be introduced to the agent's wife, unless there is an explicit operational necessity. Should the officer and the agent incidentally meet at a social event (e.g., a reception) attended by the agent's family, the acquaintance and any perceived closeness must be limited strictly to the context of the encounter and must not develop beyond that.

Occasionally, some agents may offer the intelligence officer favors of a personal nature. For instance, an agent who frequently travels to Switzerland on behalf of his company might offer to bring the officer items that are difficult to obtain in the officer's country of assignment. Another agent working at a women's clothing factory might offer to procure dresses or fabrics for the officer's wife at favorable prices. Such proposals must be politely and tactfully declined.

The intelligence officer must strictly adhere to the rule of never using the agent's services for personal benefit. If the officer accepts favors from the agent, he will begin to feel beholden to the agent and, as a result, will find it more difficult to remain demanding and objective in their professional relationship.

The intelligence officer must not accept gifts from the agent, except in those rare cases where refusal would result in the agent taking offense due to local customs or established norms within the country. In such instances, the intelligence officer must inform the station about the gift received. If the station determines it appropriate, the officer may retain the gift; otherwise, he must transfer it to the station. If the officer is unable to reject the gift outright, it is advisable to respond with a counter-gift of equivalent value, in order to avoid the impression that the officer remains indebted to the agent.

During covert meetings held in restaurants or similar venues, the officer must not allow the encounter to lose its professional character. Nevertheless, in the practical experience of intelligence work, there have been instances when the officer, especially during meals or under the influence of alcohol, became overly relaxed and began sharing personal opinions or confidences, assuming that the agent relationship was solid. This is fundamentally unacceptable and entirely unprofessional. Such conduct does not foster trust or improve relations with the agent; on the contrary, it encourages familiarity, undisciplined behavior, and loss of control.

Consumption of alcoholic beverages by the intelligence officer during meetings with agents is categorically prohibited. It inevitably undermines the officer's authority and capacity to control the situation.

If personal elements begin to intrude upon the officer's relationship with the agent, even minor ones, then gradually, and imperceptibly for the officer, the entire working basis of the relationship begins to erode. The officer ceases to be a genuine handler, and emotional closeness begins to take root, diminishing the officer's ability to be demanding. This leads to a lack of objectivity in assessing the agent's performance.

Thus, the strictly professional nature of the relationship between the intelligence officer and the agent is the principal condition for successful agent operations. To a large extent, it also determines the officer's ability to meet all the other operational requirements imposed on his conduct with the agent.

2. The Supervisory Role of the Intelligence Officer

The intelligence officer must ensure not only that his

relationship with the agent maintains a strictly professional and clandestine character, but also that his supervisory role in that relationship remains unambiguously his own. The agent must perceive the officer as a knowledgeable and demanding superior, one from whom much can be learned, but whose tasks and instructions are to be followed without question. Earning this attitude from the agent is by no means an easy task. The officer must clearly understand that authority is not gained through administrative posturing, indulgence, or domineering behavior.

To fully embody the role of mentor and supervisor, the intelligence officer must himself possess advanced political training, a solid grounding in operational knowledge, and a thorough understanding of the operational situation in the country of assignment. In the course of his educational engagement with the agent, the officer will frequently need to explain various political questions and should be ready to discuss them when necessary. If the officer lacks competence in these areas, he will not be able to satisfy the agent's inquiries and risks losing his authority as both political mentor and superior.

Of particular importance is the officer's operational competence. The agent must see in the intelligence officer a seasoned, covert, and cautious operator, yet at the same time, a decisive and confident representative of the intelligence service, one to whom he can entrust his fate. The agent must always feel confident that if he acts in accordance with the officer's recommendations, his cooperation will remain secure. Only under such conditions will the officer be able to achieve the precise execution of his instructions.

Thorough preparation by the intelligence officer on the topics related to the agent's line of work and the issues he is tasked with addressing is of paramount importance. All too often, the officer arrives at a meeting, as the saying goes, "empty-handed", i.e., he fails to obtain from the agent the information that might be of interest to the intelligence service. This typically results from the officer's perfunctory approach to preparing for the conversation, and from his poor grasp of the subjects tied to the agent's assignments.

For example, if the agent serves as a source of political intelligence, and the officer has not been keeping up with current political events, it is unlikely that productive outcomes will follow. In such cases, the officer's meetings with the agent take on the character of routine debriefings rather than directed intelligence collection. The agent will grow bored with these meetings, become progressively less engaged, and may eventually cease his cooperation altogether.

If the agent provides technical intelligence, the officer must possess at least a basic understanding of the relevant technical issues and the context in which the agent operates, especially in relation to the tasks assigned to him. The excerpt below, taken from the central apparatus of the intelligence service, highlights this point:

"An intelligence officer must possess a working knowledge of the subject areas in which his agent is engaged. This greatly enhances operational effectiveness. When the officer is asked for additional information and can provide a prompt, informed response without having to consult the central office, he can exploit countless operational opportunities that may never recur."

Among the officers I've worked with, only one possessed a sound technical background and understood what the agent was discussing. Two others limited themselves to relaying written instructions without truly grasping the subject matter. If they had taken even a modest interest in the agent's specialty, they could have conducted his operations far more effectively.

This is not to suggest that we require the assignment of qualified engineers, though that would certainly be advantageous. Rather, at a minimum, it is reasonable to expect the intelligence officer to have read several foundational texts to acquire basic proficiency. For instance, he ought to understand the fundamentals of radio technology and radar guidance systems if he seeks to gather intelligence in those areas.

In this context, the objective is not merely for the intelligence officer to exert influence on the agent or to stimulate his activity, but rather the opposite: the agent himself should begin to articulate justified and legitimate demands for greater efficiency in the conduct of operations.

The intelligence officer must be a broadly educated individual. Agents universally value officers who demonstrate not only familiarity with their home country's culture but also an understanding of the culture of the host country. This greatly enhances the officer's credibility and standing in the eyes of the agent.

The officer's supervisory role does not mean he must disregard the agent's advice, if that advice is sound. The officer is obliged to understand the agent's views on operational matters and, if the agent deems a particular method of execution more appropriate, to consider it. Only then should the officer arrive at a decision. The officer's authority is in no way diminished by such openness to agent input.

When the officer sets high standards for the agent, discipline, precision, and accuracy, he must embody those same qualities. He must lead by example. The officer must not fail to appear at scheduled meetings, nor should he make promises to the agent that cannot be kept. Under no circumstances should the officer commit to something that he is not in a position to deliver. The officer's personal example serves as a powerful pedagogical tool, reinforcing discipline in the agent and solidifying the officer's authority.

3: Attentiveness, Sensitivity, and Objectivity of the Intelligence Officer

The officer's relationship to the agent must be strictly professional in nature, but not formalistic or emotionally detached. Indifference and rigidity on the part of the officer inevitably lead to an agent's indifference to the work, emotional detachment, and eventual withdrawal.

The intelligence officer must consistently show concern for the agent, offering timely support, both material and moral. An agent places high value on gestures of concern shown at the right moment and is motivated to perform better as a result. A tactful and attentive attitude toward the agent usually evokes in him a sense of loyalty toward the officer and the service as a whole. It reinforces his confidence that he will not be abandoned in times of trouble, awakens his desire to share personal joys and struggles with his handler, and, most importantly, increases the productivity of his intelligence work.

If the officer fails to engage with the agent personally and empathetically, then in moments of personal difficulty or crisis, the agent may choose not to seek support from the officer. Instead, he may try to resolve matters on his own, which could lead to undesirable consequences. One case illustrates the risks of inattentive handling and the results it produced.

In one country, an agent codenamed "Friend" was recruited on an ideological-political basis. He held a minor position in a ministry, yet had access to materials of intelligence value. He regarded his cooperation with the intelligence service of a socialist state as his duty and never sought compensation for his work. He diligently executed assignments, often showing initiative in his operational activity.

A year after "Friend" was recruited for cooperation with the intelligence service, his material circumstances significantly deteriorated. This was due to the birth of twins, as well as a general decline in living standards in the country.

During conversations with his handler, "Friend" hinted on several occasions, albeit indirectly, that it had become very difficult for him to survive on his current salary. However, the intelligence officer, instead of empathizing with the agent's situation and raising the question of providing material assistance, deliberately avoided such conversations. As a result, "Friend" ceased speaking to the officer about his financial hardships and began independently seeking a way out of his predicament. Eventually, he managed to find another job, within the same ministry, that paid significantly more. "Friend" soon transitioned to this new position and, at the next scheduled meeting, informed the officer of the change.

In his new role, "Friend" had access to only limited materials of interest to the intelligence service.

Thus, as a direct result of the officer's inattentive and indifferent attitude toward the agent's personal difficulties, the intelligence service lost the opportunity to regularly obtain valuable information from one of its most promising collection sources.

The officer must always strive to earn the trust of the agent, creating an atmosphere of openness and sincerity in which the agent feels no hesitation in sharing both his professional observations and personal sentiments, joys, and struggles. The officer must structure his relationship with the agent so that the latter sees in him not only a demanding supervisor, but also a sensitive and attentive mentor, willing to extend both moral and material support when needed.

A critical factor in agent handling is the officer's ability to assess the agent's work objectively. This evaluation forms the basis for the agent's dossier within the central apparatus of the intelligence service and ensures the intelligence officer is equipped with exhaustive and objective data on the agent in order to properly manage the operations of the local agent network. Furthermore, one must consider that the officer may at some point be reassigned to another country or recalled to headquarters, while the agent, remaining in place, will be handed over to another officer. In such cases, the effectiveness of the new officer's work will depend heavily on the completeness, clarity, and objectivity of the characterization passed on about the agent.

When evaluating the agent's professional performance, behavior, political reliability, and moral integrity, the intelligence officer must be fully objective. No personal emotions or feelings, for example, sympathy or resentment, should be allowed to interfere with the officer's judgment of the agent.

It often happens that an officer continues working with an agent whom he personally recruited. The agent's effectiveness, as reflected in the value of the materials he provides (in cases where the agent is a source), is in large measure an indicator of how well the agent's candidacy was selected during the recruitment phase. For this reason, some officers tend, consciously or not, to present the agent in a more favorable light than is warranted by the facts, exaggerating the agent's intelligence potential. In certain cases, officers find themselves in a situation where an agent is clearly underperforming or underutilizing his access or capabilities, yet they fail to acknowledge that this state of affairs stems from their own poor supervision. Rather than critically analyze their management of the agent, such officers attempt to shift blame onto the agent.

Whether due to excessive praise or undue criticism, such behavior on the part of the intelligence officer is erroneous and ultimately damaging. An officer who inflates or deflates assessments of an agent's intelligence value, performance, or personal qualities misleads not only local leadership, but also the central headquarters. Headquarters, relying on an inaccurate characterization, may adopt an incorrect decision regarding the agent's future handling, his intelligence tasks, or potential areas of deployment.

The intelligence officer must systematically assess the agent, provide objective evaluations of his work, and thoroughly document both the agent's strengths and weaknesses. The officer must report to the central apparatus via the KGB station on all changes in the agent's political convictions, capabilities, as well as his family and professional circumstances.

4. Constant Vigilance and Control over the Agent's Conduct and Work

An officer's vigilance is demonstrated not only by carefully studying and verifying the agent network under his management, but also by ensuring that the agent never learns anything personal about the officer, or indeed about the intelligence service at all.

In the practice of field operations, there have been cases where individual officers, "deep cover illegals" (non-official cover officers), worked with an agent over an extended period, grew overly close to him, and began to disclose elements of their personal life. They revealed biographical details, discussed their time abroad, or provided other compromising information. Such behavior on the part of the officer is a direct consequence of lost vigilance and is entirely unacceptable, regardless of any perceived trust, it inevitably causes harm.

Vigilance and caution are of even greater importance when dealing with intelligence informants of limited reliability.

When handling agents, the officer must strictly adhere to tradecraft protocols. No matter how much the officer trusts an agent, he must never disclose anything that constitutes a state secret.

During operational contact, the agent will naturally inquire about the officer's background, his family, connections, or work. The officer must avoid providing direct responses. Instead, he must skillfully redirect the conversation, maintaining secrecy while still collecting necessary information from the agent about his personal life, environment, and professional circle. Such details are essential for monitoring, guiding, and influencing the agent's behavior.

If the officer begins to indulge the agent's curiosity with superficial or vague answers, the agent will gradually withdraw and cease to fully engage. The officer must avoid conversations that touch upon his personal life. However, in order to maintain the agent's interest in sharing personal information, the officer should disclose, in a deliberate and measured way, select details about himself. The goal is to create an atmosphere of sincerity and openness, not through full disclosure, but through calculated responses to specific questions, and occasionally offering anecdotes from one's "own life" (in accordance with the officer's assigned cover story or legenda).

The officer must also display restraint and caution when discussing the agent's host country's domestic political climate or international relations. Any careless remark may inadvertently reveal insights drawn from other agent sources.

One of the fundamental conditions for working with an agent is the officer's consistent firmness, his insistence on exacting standards and strict control over the agent's behavior and performance. In practical field work, there have been instances in which the officer placed himself on an equal footing with the agent. This relationship, initially perceived as collegial, often evolved into undue familiarity. Such a mistake is most commonly made by inexperienced officers managing agents for the first time.

At times, an officer, having established good rapport with an agent, begins to lower his standards, failing to notice the agent's lapses or turning a blind eye to flaws in behavior or operational output. The officer then gradually loses control over the agent's work.

Agents will often exploit this weakening of discipline. If an agent misses a task deadline and suffers no consequences, he becomes increasingly lax and inattentive to future operational demands. Lacking supervision and consistent feedback, the agent will begin to equate his intelligence work with personal service to the officer, rather than as a duty to the intelligence service.

Eventually, this attitude leads the agent to dictate terms to the officer, restructure the operational schedule according to his preferences, and set his own conditions for completing assignments. In such a situation, the concept of effective exploitation of the agent's potential becomes moot.

Thus, a properly organized system of managing, guiding, and cultivating the agent network significantly increases the effectiveness of utilizing each intelligence asset and directly contributes to the overall success of intelligence operations.

PART VI

METHODS OF TERMINATING AGENTS

Introduction

The duration of an agent's cooperation with the intelligence service varies individually. Some agents work for many years, while others are linked to the service only for the duration required to complete specific assignments for which they were recruited. In certain cases, contact with an agent is terminated for reasons such as exposure, unreliability, moral degradation, changes in political views and withdrawal from intelligence activities, recruitment by an adversary, or defection. Work with agents may also be discontinued due to illness, old age, or loss of operational utility.

The method of termination largely depends on the reason for ending the relationship and on the agent's understanding of political and intelligence matters, specifically, how much they know about the activities of the intelligence organization and the personnel involved in the case. If it is suspected that the agent whose contact is being severed might endanger the political interests of the socialist state, appropriate countermeasures must be taken to prevent undesirable actions by the agent and to neutralize any "negative consequences."

When cooperation with an agent is terminated, the intelligence officer must also consider the personal interests of the agent. In necessary cases, material support may be provided.

1. Preparations Prior to Termination of Contact

When deciding to terminate contact with an agent, it is essential to first develop measures to ensure the security of the KGB station, the intelligence organization as a whole, and the specific personnel associated with the agent. A concrete plan must therefore be developed, which should include:

The method of severing the relationship;

- Measures to safeguard the security of the KGB Station and all individuals known to the agent;
- Surveillance or monitoring of the agent after termination of contact;
- Financial and logistical considerations arising from termination (e.g., one-time assistance payments, arranging new employment, issuing a pension).

The plan must take into account the reasons for the termination of contact - whether due to changes in the operational situation, the agent's personal circumstances, the agent's completed role or functions within the service, or the extent of the agent's knowledge of the intelligence organization.

Security measures may include:

- Discontinuation of safehouses known to the agent
- Elimination of rendezvous points
- Retrieval of operational equipment in the agent's possession (such as cameras, secret writing kits, special containers for storing classified materials)
- Withdrawal from all covert addresses or logistical facilities associated with the agent
- Seizure of all technical devices previously issued to the agent

The decision and accompanying plan must be reported to the central apparatus of the intelligence service. Only after formal approval is received may implementation proceed. Execution must begin with measures to safeguard the intelligence organization involved, and this must be done in a manner that does not arouse the agent's suspicion. Only once security is assured should the process of formally ending the relationship begin.

2. Freezing of the Agent

Freezing refers to the temporary suspension of both intelligence collection and operational interaction with an agent. During this period, all regular contact is halted. The KGB employs freezing under the following circumstances:

Operational Inaccessibility: Political or mission-related considerations preclude the agent's use at the moment, though future activation remains viable

Counterintelligence Exposure: The agent has fallen under surveillance or suspicion by enemy counterintelligence, and further contact risks exposure or compromise

Loss of Operational Capability: The agent is temporarily unable to fulfill intelligence duties (e.g., due to illness, family relocation, or logistical instability)

Freezing may also be necessary when the agent requires time to become securely embedded in a new work environment, to develop a credible cover identity (Legenda), or to normalize their circumstances to prevent external suspicion.

When the operational situation in the host country deteriorates significantly, such that clandestine communication becomes unsafe, the agent is placed into frozen status. This measure ensures that neither party risks exposure until conditions improve.

In summary, freezing is a strategic suspension of engagement, preserving the agent's future utility while mitigating present threats. Reactivation occurs only when security, access, and utility have been reestablished.

In certain cases, agents may have previously provided valuable intelligence, leading to the projection of significant operational conclusions. However, before these conclusions can be validated, intelligence services may require time to verify the accuracy and good faith of the reporting.

For instance, the agent codenamed "Senya" was supplying valuable information that appeared both conscientious and accurate. However, the KGB Station received reports that Senya was allegedly collaborating with the intelligence service of a capitalist country. To avoid provocation and possible compromise, the KGB station decided to suspend all contact and freeze the agent under a favorable pretext for six months.

During this freezing period, a counterintelligence operation was initiated to verify Senya's loyalty. The operation, which extended over several months, concluded with confirmation that the reports alleging Senya's connection to the foreign service were unfounded. As a result, contact with the agent was reestablished.

When an agent is to be placed under conservation, the intelligence officer must ensure the agent is presented with a credible rationale. If the KGB station is able to communicate the actual reasons, the officer must clearly and unequivocally explain them to the agent in a manner that avoids generating suspicion. However, if the true reasons cannot be disclosed, then the cover story (Legend) must be carefully pre-formulated, thought out in detail, and delivered in such a way as to avoid planting doubt in the agent's mind. The officer provides the agent with a guidance-based timeframe for the freeze, specifying the expected period of inactivity.

Prior to freezing, the officer must either assign the agent a final task or conduct a verification interaction, so as to reinforce the agent's belief in the authenticity and continuity of the relationship.

Where operationally justified, the officer may also assign a task to be completed during the freeze. Depending on the reason for conservation, the agent may retain the possibility of being summoned urgently for a face-to-face meeting. As a rule, any decision to resume contact must originate from the intelligence side, the initiative to restore communications must come from the KGB station, not the agent.

The agent must never independently attempt to reestablish contact with the service during the conservation period or afterward. Such initiatives would violate tradecraft and compromise operational discipline.

Prior to freezing, the intelligence officer must obtain from the agent the most recent photograph, verify the agent's residential and work addresses, as well as personal and professional telephone numbers. The officer should also collect up-to-date information about locations the agent frequently visits, including clubs, cinemas, theaters, parks, sports facilities, beaches, and establish the times the agent is typically home or at work. It is also necessary to record the residential and work addresses and phone numbers of close relatives and acquaintances with whom the agent maintains steady contact and who may be able to provide information on the agent's whereabouts.

All of this data may prove critical if, for any reason, a scheduled meeting to conclude the freezing process fails to take place. In such a situation, the intelligence service may have to reestablish contact through either pre-established or emergency fallback communication paths devised in advance of conservation.

If an agent who was previously frozen is to be reinstated into active operations, but the reason for conservation was never made clear to them, it is necessary to arrange verification meetings. These are typically non-substantive, short interactions in which the officer and the agent meet without discussing operational matters and simply confirm contact and mutual recognition. The frequency of such meetings is determined by the KGB station based on the operational requirements and situational context.

Verification meetings are also used as status-check signals to confirm that the agent is at liberty, that there have been no changes to their personal situation, and that no external threats (e.g., counterintelligence interest) have emerged. These checks are especially important in cases where the agent has been placed in conservation due to potential compromise or suspected surveillance.

In certain cases, based on operational or administrative considerations, the intelligence service may provide the frozen agent with a monetary allowance, typically the equivalent of an average monthly salary for each month of the conservation period.

Before reestablishing contact with an agent who has been frozen, the officer must first verify that no changes have occurred in the agent's status which could obstruct renewed contact or further use of the agent in intelligence operations.

In intelligence work, such verification is especially essential if the agent was transferred to conservation due to the threat of compromise. This verification ensures the agent's continued operational viability.

3. Non-Attendance at a Meeting as a Method of Severing Contact

This method consists in the intelligence officer deliberately failing to attend a scheduled or fallback meeting with the agent, thereby unilaterally terminating the relationship. This approach becomes primarily necessary when a planned meeting with an agent or courier has become operationally hazardous, for example, if the agent has come under suspicion by counterintelligence, thus requiring the immediate severance of contact without any warning.

In cases where an agent under surveillance by counterintelligence suddenly has their contact terminated, the aim is to prevent the agent from being placed in a more dangerous situation. Continuing contact could allow the adversary to gather concrete proof of the agent's collaboration with the intelligence service. When circumstances allow, the agent should be briefed in advance on such emergency scenarios, and advised that a no-show at a scheduled meeting may serve as a signal that contact is being deliberately severed.

However, the unannounced discontinuation of meetings with a mole agent can itself provoke suspicion, particularly if the mole was attempting to lure the officer into a setup. In such cases, counterintelligence may view the absence of the officer as confirmation that the agent was indeed connected to foreign intelligence. Thus, the officer must weigh the risks and determine the safest course of action.

If the risk posed by severing contact exceeds the risk of continuing under surveillance, it may be more prudent to execute a severance under the guise of "conservation." This allows time to implement protective measures and extricate the officer or assets as needed.

Severing contact with a mole through non-attendance can

escalate matters if not handled correctly. Counterintelligence, upon observing the agent's dismissal, may conclude the agent has outlived their usefulness and coerce them into betraying classified information, exposing other agents, or engaging in false-flag operations. Therefore, the decision to terminate contact via this method must be undertaken with complete operational clarity.

The KGB station develops and implements measures aimed at localizing the operational failure and preventing any potential provocation by adversary counterintelligence. In particular, the station carefully considers the issue of the future disposition of the agent network and of the intelligence officer or agent who had been in contact with the mole.

4. Termination of Contact under a Legendized Pretext

Legendized pretexts for severing contact are employed in those cases where a unilateral termination, such as through the non-attendance method, could lead to negative consequences (such as attempts by the agent to reestablish contact, or other actions detrimental to intelligence operations). Legendized explanations may also be used in situations involving the temporary cessation of intelligence activities in a country, threats of compromise, illness, or the departure of the intelligence officer from the operational environment. In all such cases, the severance must be credibly justified. Special operational measures may be undertaken to reinforce the plausibility of the legend.

When this method is used to sever contact with a mole agent, intelligence operations aim to conceal the true nature of the severance with maximum care. To enhance the perceived credibility of the legend, an agent may even be provided with material support. Everything necessary is done to ensure that the termination appears natural and unconnected to counterintelligence concerns.

This buys valuable time to implement countermeasures and contain the damage from the breach. Most often, this method is used when contact is being severed not due to betrayal, but for reasons such as the loss of operational value, unreliability, or serious tradecraft negligence.

5. Open Termination of Contact with the Agent

An open termination of contact with an agent consists of informing the agent directly that his relationship with the intelligence service is being ended, along with an explanation of the reason for the severance. This method is applied primarily when there is no concern about operational risks or complications. It may be used in relation to agents whose contact is being terminated due to loss of operational value, or due to significant personal changes that render continued cooperation impossible, such as exposure, psychological instability, or ideological drift.

This method may also be used when termination is required due to the agent's poor performance or misconduct, such as falsification of intelligence materials or attempts at extortion. In such cases, the intelligence officer must warn the agent of the possible consequences should he attempt actions detrimental to the service.

If the agent had previously rendered valuable service, he may, depending on circumstances, be granted material assistance. However, the open termination of contact involving a known mole, even with a stated reason, is highly uncommon. In such situations, the intelligence officer may explain that the agent is under suspicion by counterintelligence, and that continuation of contact may be exploited by the adversary to fabricate accusations or provocations against the intelligence officer.

6. Transfer of the Agent to Another Country

Intelligence may also use the method of relocating the agent to another country as a means of terminating operational contact. For example, the agent may be transferred out of the country where he had previously operated, either to the country overseeing the operation, or to another socialist state. Alternatively, the agent may be sent to a capitalist country.

This method is applied in cases where the agent can no longer be retained in the current country of operations, and where only relocation can definitively and safely sever the relationship.

The removal of an agent to another country may be necessitated by his exposure, threat of arrest by counterintelligence services, or the danger of retaliation from reactionary or anti-Soviet organizations against which he had worked.

An agent may also be transferred abroad if his continued presence in the target country is deemed operationally unsafe due to his extensive knowledge of intelligence procedures. Such a transfer may be conducted with the agent's awareness, particularly if he had demonstrated loyalty and conscientious collaboration. In other instances, however, the agent must not be made aware that the true purpose of his relocation is to sever ties with him permanently.

In some cases, intelligence may have only limited time to execute such operations. Successful execution requires full mobilization of the KGB station and support from Center. Agents known or suspected to be moles or traitors are typically transferred abroad with the intent of terminating their activities and neutralizing the threat of compromise to Soviet operations. These operations also serve to identify foreign intelligence and counterintelligence activities, or the roles of particular hostile organizations.

When transferring a compromised or treasonous agent to a socialist country, the transfer is managed through agent-operational deception, designed to convince the adversary's counterintelligence service, as well as the mole, that the Soviet service still trusts the agent and that the relocation is motivated solely by operational considerations.

Depending on the reason for transfer, the agent's legal status, and the available options, the relocation may be conducted either legally, with all necessary documentation and formalities, or clandestinely, by covert border crossing or use of false identity papers.

Printed in Dunstable, United Kingdom

66325689R00087